Naturalizing God?

Naturalizing God?
A Critical Evaluation of Religious Naturalism

Mikael Leidenhag

Published by State University of New York Press, Albany

© 2021 State University of New York

All rights reserved

Printed in the United States of America

No part of this book may be used or reproduced in any manner whatsoever without written permission. No part of this book may be stored in a retrieval system or transmitted in any form or by any means including electronic, electrostatic, magnetic tape, mechanical, photocopying, recording, or otherwise without the prior permission in writing of the publisher.

For information, contact State University of New York Press, Albany, NY
www.sunypress.edu

Library of Congress Cataloging-in-Publication Data

Name: Leidenhag, Mikael, 1986– author.
Title: Naturalizing God? : a critical evaluation of religious naturalism / Mikael Leidenhag.
Description: Albany : State University of New York, [2021] | Includes bibliographical references and index.
Identifiers: LCCN 2021010434 (print) | LCCN 2021010435 (ebook) | ISBN 9781438484419 (hardcover : alk. paper) | ISBN 9781438484402 (pbk. : alk. paper) | ISBN 9781438484426 (ebook)
Subjects: LCSH: Naturalism—Religious aspects.
Classification: LCC BL65.N35 L45 2021 (print) | LCC BL65.N35 (ebook) | DDC 210—dc23
LC record available at https://lccn.loc.gov/2021010434
LC ebook record available at https://lccn.loc.gov/2021010435

10 9 8 7 6 5 4 3 2 1

To my mother, Carina Leidenhag (1954–2016)

Contents

Acknowledgments	xi
Abbreviations	xiii
Introduction	1
The Religious Naturalists to Be Discussed	1
Current Research on Religious Naturalism	3
Metaphysical Grounding Problems	7
Chapter Overview	8
1 Explicating Religious Naturalism	11
Introducing Religious Naturalism	11
The Nature of Naturalism	18
Religious Aspects of Reality	21
Religious Naturalism and Traditional Religion	25
The Function of Religious Language	31
Conclusions	38
2 What Is Naturalistic about Religious Naturalism?	41
The Central Pillars of Naturalism	41
Monistic Naturalism	46
Pluralistic Naturalism	49
Summary of Monistic Naturalism and Pluralistic Naturalism	56
Conclusions	57
3 The Metaphysical Grounding Problems of Monism and Pluralism	59
Monistic Naturalism and the Issue of Antireductionism	60
Pluralistic Naturalism and Emergence Theory	67

	Some Grounding Problems for Pluralistic Naturalism	70
	Conclusions	80
4	The Religious Dimension of Religious Naturalism	83
	Religious Realism in Rue and Crosby	83
	Religious Antirealism in Hardwick and Drees	89
	Understanding Pragmatic Religious Realism	94
	Conclusions	106
5	The Problem of Religious Discourse in Religious Naturalism	109
	The Tension between Physicalism and Christian Faith	109
	Limit-Questions and the Status of Naturalism	114
	Pragmatic Realism	116
	Functional Religion and Theological Realism	120
	Objectivist Religion and the Problem of Evil	125
	Conclusions	134
6	Alternative Ontology 1: Naturalistic Options	137
	Liberal Naturalism	137
	Agnostic Naturalism	146
	Pragmatic Naturalism	151
	Conclusions	156
7	Alternative Ontology 2: Theistic Options	159
	Background to Panentheism	160
	Why Panentheism Entails Dualism	163
	The Positive Status of Panentheism and the Question of Religious Naturalism	170
	God and Values: A Proposal by Fiona Ellis	171
	Conclusions	179
8	Alternative Ontology 3: Panpsychism	181
	Panpsychism Today	182
	Three Arguments for Panpsychism	184
	Between Strong and Weak Panpsychism	187
	The Panpsychist Dimension of Emergence Theory	189
	Metaphysical Objections to Panpsychism	199
	The Religious Relevance of Panpsychism	206
	Conclusions	216

9 Concluding Remarks and Looking Ahead	219
Panpsychism and Teleology	220
Panpsychism and Eco-Ethics	222
Subjectivity and the Promise of Panpsychism	224
Notes	227
References	249
Name Index	261
Subject Index	265

Acknowledgments

Writing this book has been both a challenging and a rewarding experience, and I have many people to thank for supporting and inspiring me throughout this time. I owe special thanks to my PhD supervisors, Mikael Stenmark and Ulf Zackariasson, for their support, valuable feedback, and advice during this project.

I would also like to thank the research seminar in Philosophy of Religion at Uppsala University for reading and commenting on various chapters and articles. Those who deserve special mention include Johan Eddebo, Karin Johannesson, Francis Jonbäck, Maria Klasson Sundin, Lotta Knutsson Bråkenhielm, Oliver Li, and Mikael Sörhuus. Niels Henrik Gregersen also deserves special thanks for examining the PhD thesis on which this book is based. During my PhD studies I had the opportunity and privilege to spend two semesters doing research at Princeton Theological Seminary. I would like to thank both Wentzel van Huyssteen and Gordon Graham for supervising me during my time in Princeton, and for including me in such an inspiring research environment.

I am also grateful to *Forum Philosophicum* for letting me reprint my article "Is Panentheism Naturalistic? How Panentheistic Conceptions of Divine Action Imply Dualism," in *Forum Philosophicum* 19, no. 2 (2014): 209–225. This article appears in a modified form in chapter 7.

Finally, and most importantly, I would like to express my gratitude to my lovely wife, Joanna. She has not only supported me emotionally during stressful times but also read through this book, offered valuable critique, and helped me improve it. Thank you for sharing this journey with me.

Mikael Leidenhag
St. Andrews, July 2020

Abbreviations

Monistic Naturalism

MON1 The natural world is all there is, and all entities are made up of the same constituents.

MON2 If X exists, X is either something material or a property of matter.

MON3 The ultimate cause of things is natural, nonpersonal, nonmental, and nonintentional.

MON4 Higher-level-happening Y corresponds to lower-level-happening X.

MEN1 Physics offers the best description of reality.

MEN2 All truth is determined at the level of physics.

MEN3 S should believe proposition X only if X is well supported or if X has been experimentally verified.

MSN1 Normative statements X, Y, and Z are reducible to the vocabulary of science.

MSN2 Normative statements X, Y, and Z are cognitively meaningless.

Pluralistic Naturalism

PON1 Reality consists of a hierarchy of higher and lower levels.

PON2 Higher-level Y has emerged from lower-level X.

PON3	Higher-level *Y* cannot be reduced to or be replaced by lower-level *X*.
PON4	We are epistemologically unable to deduce higher-level entities/properties from low-level physical laws.
PON5	Higher-level phenomena *Y* can exert causal efficacy on its constituent parts.
PEN1	There are other disciplines than science capable of producing knowledge about reality.
PEN2	There are meaningful claims that fall outside the boundary of science.
PSN1	Teleological language is irreducible and emergent with respect to physics.
PSN2	Some explanations may require vocabularies that go beyond physics, or the natural sciences in general.

Introduction

Once almost extinct, religious naturalism is now making a return to the intellectual scene and is gaining renewed scholarly attention. Jerome A. Stone, in his book *Religious Naturalism Today—The Rebirth of a Forgotten Alternative*, explores this neglected option through an analysis that begins with the twentieth-century figure George Santayana and the British emergence theorist Samuel Alexander. Stone provides a comprehensive historical survey of religious naturalism, bringing out the similarities and differences between a variety of thinkers, and shows how this form of naturalism has become a viable religious option.

Religious naturalists seek to develop a middle path between scientific reductionism and supernaturalism, neither of which is seen as tenable. For a religious naturalist, nature provides the definitive foundation for a religious way of life. Nature is both metaphysically and religiously ultimate, meaning that there is nothing above and beyond the natural domain. In addition, several religious naturalists maintain that traditional religion, with its distinction between God and nature, is uniquely responsible for the current ecological crisis. The idea here is that traditional religions have failed to appreciate the intrinsic value of nature. Several religious naturalists therefore propose new images of God that are closely related to the workings of nature. Such new images, they argue, can inspire people to adopt more beneficial attitudes toward the natural world and its ecosystems.

My aim is to critically evaluate religious naturalism as a position in the dialogue between science and religion, and to see what possibilities there are for developing and moving this perspective forward.

The Religious Naturalists to Be Discussed

The religious naturalists discussed and critically evaluated in this book construe both naturalism and religion in a number of different ways. In this discussion we find expressions of what is commonly referred to as hard naturalism and soft naturalism.[1] Some opt for a more restrictive (harder) form of naturalism, suggesting that science is the primary, if not the only, source of knowledge. Science therefore sets the boundaries for what can exist. On this more restrictive side of the spectrum are Willem B. Drees and Charley D. Hardwick, whom I will analyze. Both define naturalism in more materialist and physicalist terms, and both see naturalism as an approach continuous with, and intimately linked to, the empirical sciences. Hence, they both understand naturalism to be an approach that takes the methodologies and discoveries of science very seriously.

On the soft side of the spectrum we find those naturalists who maintain that reality is layered, and that consciousness, values, and meaning are fully natural yet irreducible features of reality. There is nothing supernatural about such phenomena, but they transcend the boundaries of empirical inquiry. This form of naturalism is found in the writings of Donald Crosby, Ursula Goodenough, Stuart Kauffman, Gordon Kaufman, Karl Peters, and Loyal Rue. These thinkers will also be analyzed. Despite some significant differences, it should be noted that hard and soft naturalists both agree that we live in a fully natural world, devoid of supernatural and extranatural beings. Indeed, as Crosby puts it, "The antithesis of religious naturalism is any kind of supernaturalism."[2]

Different forms of naturalism bring with them and enable a diversity of ways to religiously and spiritually engage with nature. There are those who take nature itself to be religiously significant, and those who do not. Hardwick, on his physicalist view, thinks that there is nothing in nature that is religiously significant or that calls for religious attitudes of awe and wonder. Yet he maintains that humans can experience "events of grace," unexpected events that offer the possibility of self-transformation so as to become existentially open to the future. Drees, whose position lies somewhere between religious naturalism and religious agnosticism, places the religious significance on the limit-questions pertaining to the ultimate issues of human existence. Examples of such questions are "Why do we exist?" and "Why is there something rather than nothing?" For Drees, limit-questions create space for a religious interpretation of reality.

Those who subscribe to a soft version of naturalism maintain, contrary to Hardwick, that nature, either as a whole or certain aspects of it, can provide a sound religious foundation for naturalism. Crosby, who prefers to call his perspective "Religion of Nature" or simply "Naturism," maintains that nature as a whole is to be regarded as humanity's religious object, worthy of the awe and reverence traditionally directed toward God. Others, such as Peters, Kaufman, and Kauffman, instead take particular aspects of nature to be religiously significant. God, on these views, is identified with the biological processes of nature, which give rise to unexpected and novel phenomena. The emergent process, which is celebrated as God, involves both goods and evils, life and death. In this way, God is constructively imagined as the creativity in the universe. Given the unexpectedness and intrinsic unpredictability of this creative process, it is said to invite a sense of mystery and awe for the hidden dimensions of nature.

We can see here that different forms of naturalism generate different views regarding the religious significance and potentiality of nature. It will also be seen that some religious naturalists employ God-talk, while others refrain from it. Those who focus on the concept of "creativity" often associate this with God, and hence rely on traditional religious language. Others, like Drees, Crosby, and Rue, do not employ God-language in their naturalistic interpretations of religion. Crosby instead talks about Nature with a capital "N" as the ultimate religious object. Likewise, Rue, who focuses on the "Epic of Evolution," constructs a form of religious naturalism that seeks to establish harmony between humanity and Nature.

Religious naturalists who focus their religious attention on nature tend also to express ecotheological views. Thinkers such as Peters, Kaufman, Kauffman, and Rue suggest that traditional and dualistic conceptions of God have facilitated an underappreciation of the natural world. However, rather than rejecting religion, they seek to engage with religious discourse constructively and to formulate new images of God, the Sacred, and the Divine consistent with naturalism. The hope is to motivate religious believers to act in an ecologically responsible manner. Ecological awareness is not a determinative issue for *all* religious naturalists. But, for a significant number, it seems to form a central pillar in their proposal and is therefore given extra consideration in this book.

Current Research on Religious Naturalism

Through an overview of some of the significant research contributions on this emerging perspective, we can gain a better understanding of what the term "religious naturalism" connotes. It is important to point out from the start that religious naturalism is a pluralistic perspective and it cannot be reduced to one standpoint, or a single principle or belief. Indeed, to more fully understand this perspective, it is better to view it in terms of *family resemblance*. As Willem Drees writes, "Religious naturalism is an umbrella term which covers a variety of dialects, of which some are revisionary articulations of existing traditions whereas others may be more purely naturalistic religions indebted almost exclusively to the sciences. There is family resemblance, with affinities and disagreements, not unity."[3] This family resemblance and plurality is clearly reflected in the research that has been done so far on this religious option. I have already mentioned Jerome Stone's extensive contribution to this research area. In his seminal book *Religious Naturalism Today*, as well as in several of his articles, Stone has brought out the distinctiveness of this naturalistic religiosity. Religious naturalism "is the attitude and belief that there are religious aspects of the world which can be conceived within a naturalistic framework."[4] This perspective, by virtue of being naturalistic and religious, entails a negative and a positive side. Negatively and naturalistically, it excludes the idea of an "ontologically distinct and superior realm."[5] Positively, this perspective maintains that our religious focus should be "on the events and processes of this world to provide what degree of explanation and meaning are possible to this life."[6] This definition offered by Stone captures well two central belief-components for religious naturalists.

Another helpful overview comes from Michael Cavanaugh and his article "What Is Religious Naturalism? A Preliminary Report of an Ongoing Conversation." Cavanaugh defines religious naturalism as "a belief in the natural order as understood by ongoing *scientific investigation*, supported by a strong and positive feeling about the wonder and efficacy of that natural order."[7] Cavanaugh adds an important dimension to this definition, namely the strong reliance on *science* in order to understand the structures and workings of nature. This is, I believe, a defining feature of contemporary religious naturalism. Science functions in two ways within this naturalistic framework: as a critique of traditional expressions of religion (supernaturalism, in particular), and as a way to offer a description of reality that can elicit responses of awe and wonder. The "epic of evolution" serves as a religious metanarrative for many spiritually inclined naturalists. Science for these naturalists is not merely a fact-producing enterprise; it can also shed light on (while not being able to offer conclusive answers to) the ultimate questions regarding purpose, values, and human existence.

In the article "Religious Naturalism and Science," Willem Drees has carefully outlined several ways of positively construing the relationship between religion and naturalism. Drees suggests that speaking of "*religious* naturalism may thus be justified if the attitudes and responses are sufficiently religious."[8] Drees introduces a variety of thinkers that he suggests represent contemporary religious naturalism. He mentions Gordon Kaufman's Christian interpretation of naturalism, and Charley Hardwick's Christian physicalism, but also those naturalists that seek to develop a religious alternative independent of existing traditions, such as Loyal Rue, Ursula Goodenough, Jerome Stone, and Donald Crosby.[9] Drees recognizes the diversity within religious naturalism but concludes that the common core of this emerging movement is its ambition to maintain a religious attitude that is consistent with science, naturalistically conceived.

Donald Crosby has provided a survey of religious naturalists in his article "Religious Naturalism," which appears in *The Routledge Companion to Philosophy of Religion*. Similar to both Stone and Cavanaugh, Crosby defines religious naturalism as the antithesis of supernaturalism and ontological theism. Crosby, similar to Cavanaugh and Drees, stresses the importance of the natural sciences for this worldview: "Religious naturalists take seriously the *methods and findings of the natural sciences*. They seek to develop religious outlooks consistent with these methods and findings, and to avoid the sorts of conflict between science and religion that have plagued religious traditions of the West in the past. They are also religiously inspired by the discoveries of science and especially by scientific descriptions of cosmic, terrestrial, and biological evolution."[10] As we will also see, several thinkers that I discuss emphasize the ethical potential and relevance of religious naturalism for the current ecological crisis. In *The Promise of Religious Naturalism*, Michael S. Hogue situates this religious approach within a posttraditional setting. Hogue, through his appreciative criticism, maintains that the primary aim of contemporary religious naturalism (focusing on Jerome Stone, Loyal Rue, Donald Crosby, and Ursula Goodenough) is to develop a morally and religiously significant response to the perils of the ecological situation.[11] This ecological crisis, Hogue argues, is "morally degrading and spiritually and religiously deadening."[12] It is indeed a problem not just for humanity but for the "whole future of life."[13] Hogue further claims that the well-being of nature should be understood as a *religious* concern, as the crisis of ecology is functionally equivalent to a spiritual crisis. Thus with regard to contemporary religious naturalism Hogue concludes that the ecological emphasis is intrinsically intertwined with the religious goal.

The motivation to develop an adequate religious response to the environmental situation is of primary concern for several religious naturalists, but it should be pointed out that this is not a major concern for all naturalists discussed herein. For Charley Hardwick and Willem Drees, the ecological crisis is not a determinative factor in their naturalistic reinterpretations of religious discourse. However, for Karl Peters, Gordon Kaufman, Stuart Kauffman, Ursula Goodenough, and Loyal Rue, the primary aim of religious naturalism is to develop new images of God and the Sacred that can motivate people to act in an ecologically sensitive manner. I will therefore assess those forms of religious naturalism that take into account the ecological crisis, such that religious images are considered on the basis of their ecological adequacy. I will also critically discuss those forms of religious naturalism that take, for example, the harmony between religion and science to be of primary concern.

As stated above, "religious naturalism" is an umbrella term that covers a wide range of perspectives and beliefs. Indeed, not all thinkers to be discussed herein are comfortable with the label of "religious naturalism." Stone has discussed this issue. He notes that Hardwick, Goodenough, and Crosby have identified their views as forms or expressions of religious naturalism.[14] Hardwick refers to his view as "naturalistic theism."[15] However, he also describes his perspective as a form of religious naturalism, but one that "is constrained by physicalism."[16] Goodenough freely employs the concept of "religious naturalism" to describe her approach to religion, and the idea that science "can call forth appealing and abiding religious responses."[17] Crosby prefers to label his approach "religion of nature," yet he acknowledges his view to be "one of at least four general categories of religious naturalism."[18]

Karl Peters does not use the term "religious naturalism" to describe his perspective. Seeking to reform Christian faith through a naturalistic framework, Peters chooses to call his perspective "Christian naturalism" and "naturalistic theism."[19] This, I suggest, should not be understood as a rejection of religious naturalism. Instead, he embraces the term "Christian naturalism" because he approaches naturalism from a specific tradition. Yet, given that Peters seeks to modify religion according to a naturalistic framework, and given his strong belief in science's ability to uncover the sacredness of nature, it seems appropriate to conclude that his perspective belongs to the general category of religious naturalism.[20] The same thing should be said about Gordon Kaufman, who labels his view "biohistorical naturalism." This term is powerful for Kaufman as it helps to stress our embeddedness in evolutionary history and the becoming of the universe.[21] Nevertheless, like other religious naturalists, Kaufman stresses the need for religion to change in light of the advancements in science, and the ability of science to point us toward the religious dimensions of nature.

Drees explicitly affirms a naturalistic understanding of reality, but he is rather hesitant when it comes to adopting the term "religious naturalism" for his view. He writes, "Am I a religious naturalist? Others have used that label for me. I am not sure that I like the label, as it seems to constrain, whereas I want to explore. . . . Even if I am not sure whether I am a religious naturalist, I am most interested in understanding what religious naturalism might mean, may become, and will offer."[22] As mentioned earlier, Drees's perspective can be located somewhere between religious naturalism and religious agnosticism. I maintain, though, that Drees, in the way that he seeks to naturalistically reconceive religion, can be considered a religious naturalist.

Stuart Kauffman is an interesting case in this debate, as he employs neither the term "religious naturalism" nor the general term "naturalism." Kauffman's stance on ontology is overall less clear.[23] However, he denies the existence of a creator God, and claims that whatever exists must be compatible with the laws of physics.[24] He furthermore affirms strong emergence, which suggests that whatever has emerged in nature has its origin in something physical.[25] In this way, Kauffman seems to affirm a form of naturalism. Moreover, as he seeks to naturalize God and reinvent God as the Sacred creativity in the universe, I think it is fair to consider him a religious naturalist.

Metaphysical Grounding Problems

We will throughout this book discuss several metaphysical problems that challenge the plausibility of a religiously naturalistic outlook on reality. Broadly speaking, metaphysics is in the business of bringing into light and analyzing the variety of ontological assumptions that we employ, consciously or unconsciously, in philosophy, theology, the natural sciences, and everyday life.[26] Metaphysics is a way of finding out if a way of talking is compatible with the available ontological resources of a metaphysical system. For example, we might want to investigate if a certain form of moral language is compatible with a particular ontology. Is it coherent for a naturalist to employ notions such as "right" and "wrong" given her ontology? Is it possible for a naturalist, who maintains that all of reality is natural, to hold moral and ethical properties to be real? Is it possible for a Christian to employ agential language regarding human creatures if God is omniscient? That is, does a theistic framework rule out free will for human creatures and therefore agential language?

I suggest that a conflict between a set of beliefs and a metaphysical framework can be referred to as a *metaphysical grounding problem*. That is, it is not possible to ground one's beliefs or vocabulary within the relevant ontological framework. I will explore and analyze some of the potential metaphysical grounding problems facing religious naturalism. In chapter 2, I outline two versions of naturalism, referred to as monistic and pluralistic naturalism. In outlining the different ontological, epistemological, and semantic commitments, I seek to show and analyze potential grounding problems for religious naturalists. I will put both monistic and pluralistic naturalism to the test and evaluate how successful they are when it comes to grounding higher-level language within their framework.

Chapter Overview

I will provide a critical evaluation of religious naturalism as a position in the dialogue between science and religion. While this book provides an overview of the most important issues and central arguments within religious naturalism, it should be stated from the start that I will not be able to discuss each one of these issues or arguments in full detail.

Chapter 1 sets the stage for the book by describing how religious naturalists view the authority of science and the ways in which they construe naturalism as compatible with a religious appreciation of nature. This chapter also explores the relationship between religious naturalism and traditional religion, and outlines a constructive proposal for how to understand the function of religious language within this emerging perspective.

As the term "naturalism" is ambiguous, and used in multiple ways by religious naturalists, chapters 2 and 3 probe more deeply into this concept. It will be seen that some religious naturalists lean toward a softer version of naturalism, while others venture into more reductive outlooks. These positions are described under the headings of pluralistic and monistic naturalism. It is argued that both encounter significant metaphysical grounding problems, which consequently puts religious naturalism on a shaky foundation.

How should we understand the religious component of religious naturalism? Chapter 4 seeks to respond to this question by outlining both realistic and antirealistic approaches to religious discourse and the nature of religion. As will be seen, a pragmatic version of religious realism seems to be the dominant position among proponents of religious naturalism. Chapter

5 goes on to critically evaluate the three approaches to religious discourse and identifies significant problems in each of them.

In light of the problems of both monistic naturalism and pluralistic naturalism, this book evaluates alternative ontological frameworks for moving religious naturalism forward. The first alternative ontology is discussed in chapter 6, and centers on alternative ways of understanding philosophical naturalism. Both liberal naturalism and agnostic naturalism are critiqued, because it remains unclear from these perspectives why we should prefer naturalism to other ontological frameworks. Pragmatic naturalism is also deemed unsuccessful not only as it fails to appreciate the seriousness of some philosophical problems, but also because of the way that it undermines the authority of science and naturalism as such. I therefore conclude that none of these alternative naturalisms can help religious naturalism in moving forward.

While some propose new formulations of naturalism to ease the seeming tension between science and religion, others focus more directly on finding new conceptions of God consistent with the findings of science. Chapter 7 evaluates one such attempt, namely panentheism. The conclusion of this chapter is that panentheism, defined in conjunction with either emergence theory or against the background of process philosophy, implies dualism. The panentheistic alternative seems unsuccessful, and thus proponents of religious naturalism should look elsewhere. Fiona Ellis's attempt to combine Christian theism with naturalism is also evaluated. This approach, in a similar way to panentheism, seems unable to explicate God's action within nature.

Having looked at various alternative naturalistic ontologies and panentheistic frameworks, this book offers panpsychism as the final and most promising framework for religious naturalism. Chapter 8 explores the metaphysical, religious, and ecological benefits of panpsychism, and suggests that it should be seriously considered by proponents of religious naturalism. In this chapter I further argue that emergence theory and panpsychism are not mutually exclusive but reciprocally enriching. I also show how some religious naturalists already seem to be expressing panpsychist ideas. Chapter 9 concludes this book by further exploring the promises of panpsychism for the science-religion dialogue.

1

Explicating Religious Naturalism

This chapter seeks to explicate the nature of religious naturalism. What is it that makes this perspective both *naturalistic* and *religious*? How should we understand the motivation behind this worldview? Where do religious naturalists place science on the epistemological ladder? Should we consider religious naturalism to be a realistic or antirealistic perspective in the science-religion dialogue? And what is the relationship between this emerging worldview and traditional religions? These are some of the central questions that will be addressed in this chapter.

Introducing Religious Naturalism

Religious naturalism is a perspective in which the natural world is viewed as being metaphysically ultimate but at the same time religiously adequate or at least highly relevant. Meaning, transcendence, grace, the Sacred, and what is Good can and should be located in the natural order, or some aspect of the physical domain.[1] Religious naturalists believe that certain aspects of reality are religiously significant and can be appreciated within a naturalistic framework.[2] As such, religious naturalism can in many ways be construed as a middle path in the science-religion dialogue. On one side, religious naturalists seek to avoid every form of supernaturalism that postulates an ontologically distinct being that is in some way separate from us and the universe. On the other side, they resist any type of atheism that depicts the universe and the whole of reality as being meaningless or metaphysically insignificant.[3] They also reject those forms of militant atheism that have been put forward in the public sphere by Richard Dawkins, Sam Harris, and Daniel Dennett. Jerome Stone therefore considers religious naturalism to be a third alternative to the dichotomy between supernaturalistic theism and full-blown secular humanism.[4]

For the majority of religious naturalists, the desire to locate purpose, transcendence, and meaning in nature is connected to an ecological awareness and ambition to formulate an ecologically sensitive religious alternative. The ecological crisis is currently being manifested in climate change, an increasing starvation rate, pollution, a reduced ozone layer, infectious diseases, the energy crisis, depletion of the earth's fresh water, and so on. This crisis is, according to many, "one of the most urgent moral and political and religious challenges of our time, regardless of one's political leanings or religious affiliation or social location."[5] Many religious naturalists argue that we must, if we want to respond effectively to the ecological crisis, propose new images of "God," "Spirit," "the Sacred," and "Transcendence," which can justify an eco-ethical view of nature as possessing absolute value. Some proponents of religious naturalism have likened their position to a form of "ecotheology."[6]

Broadly speaking, the issues outlined above characterize what is known as religious naturalism. In the remainder of this chapter I will provide an overview of the central ideas that define religious naturalism as a position in the science-religion dialogue.

Historical Predecessors

Jerome A. Stone, a forerunner in the scholarly debate on religious naturalism, notes that this worldview is experiencing a revival.[7] He seeks to contextualize religious interpretations of naturalism, highlight some of the significant differences between religious naturalists, and further bring out several important factors that can explain how and why this perspective is making a return.

Stone notes that Baruch Spinoza's (1632–1677) pantheistic view of God bears some resemblance to how God is immanently conceived by several religious naturalists. Spinoza construed the relationship between God and nature by postulating two aspects of nature: "First, there is the active, productive aspect of the universe—God and his attributes, from which all else follows. This is what Spinoza, employing the same terms he used in the *Short Treatise*, calls *Natura naturans*, 'naturing Nature.' Strictly speaking, this is identical with God. The other aspect of the universe is that which is produced and sustained by the active aspect, *Natura naturata*, 'natured

Nature.'"[8] However, I would add that unlike religious naturalists, Spinoza did not seek to *naturalize* God. God, in the pantheistic vision, was a real aspect of nature, and the workings of nature "followed immediately from God's natures."[9] Spinoza was likely not a reductive pantheist in the sense of completely removing God's transcendence. Moreover, Spinoza, unlike religious naturalists, argued that nature is not sacred and so cannot be a religious object. On Spinoza's view, "There is nothing holy or sacred about Nature, and it is certainly not the object of a religious experience. Instead, one should strive to understand God or Nature, with the kind of adequate or clear and distinct intellectual knowledge that reveals Nature's most important truths and shows how everything depends essentially and existentially on higher natural causes."[10] This is an important difference between Spinoza and some contemporary religious naturalists, as Spinoza strongly maintained that religious awe before nature/God would give rise to superstition and submission to religious authorities. There are some similarities between Spinoza's pantheism and religious naturalism, but there are also important differences that set them apart.

Stone identifies three significant figures that have influenced contemporary religious naturalism. These are George Santayana (1863–1952), Henry Nelson Wieman (1884–1975), and John Dewey (1859–1952).[11] Santayana, like contemporary religious naturalists, emphasizes the practical importance of religious beliefs in everyday life. Yet the Spanish-born philosopher and poet ultimately found supernatural and dualistic beliefs to be ontologically false. Instead of viewing religion as a collection of ontological assertions about ultimate reality, Santayana adopted a functionalist stance with regard to religion. As Willard E. Arnett remarks, for Santayana religion has a *poetic function*. Arnett writes, "Whatever ideas, ideals, or figments may be expressed in religion—and poetry—are thoroughly human and must be understood in terms of their genesis and function, even if they cannot be accepted as indicative of the nature of the universe outside man's experience."[12] As religion can no longer be expressed in traditional and otherworldly language, Santayana proposes this functionalist and at the same time immanent conception of religion. Hence, religious ideas and concepts must refer to *this* world and the challenges facing humanity. This shift in emphasis fits rather well with many of the theoretical guiding points of religious naturalism.

A comparable view can be found in the writings of Wieman (who is also discussed in chapter 5). Wieman proposes the notion of God as the creativity of nature and the source of human good. In a similar vein to Santayana, Wieman suggests that we must ignore "the transcendental affirmation in the Jewish Christian tradition of a creative god who not only works in history but resides beyond history."[13] Wieman suggests instead that "the only creative God we recognize is the creative event itself."[14] This focus on the creativity of the natural order is central to the proposals of Karl Peters, Gordon Kaufman, Stuart Kauffman, and to some extent Donald Crosby.

The functionalism expressed by Santayana, and to a degree by Wieman, can also be found in the pragmatist thinker John Dewey. Indeed, as Willem Drees contends, religious naturalism and pragmatism overlap with regard to interests and aims. They are both concerned with life here and now and are characterized not by adherence to a supreme authority but by "self-reliance and taking responsibility for one's thinking."[15] Yet self-reliance does not, according to Drees, mean that our achievements can be thought of as isolated from history and the hard work of earlier generations. Rather it invites "gratitude to earlier generations and the whole of nature, the sources of our existence," a "gratitude that is honoured not by receptivity alone, but by moving on, by further explorations."[16]

The pragmatic religious naturalism of John Dewey, as primarily expressed in *A Common Faith*, suggests that "the religious" could be separated from actual religions and religious institutions. As Sami Pihlström explains, for Dewey "the religious aspects of experience can be appreciated without metaphysical commitments to anything supernatural."[17] Hence, by separating the "the religious" from the ontological, Dewey, much like contemporary religious naturalists, raises the issue of realism with regard to religious utterances. Pihlström explains further the importance of the distinction between "the religious" and "religions." He writes, "Dewey is about to tell us what is 'genuinely religious'—apparently in contrast to what is pseudo-religious or superstitious."[18]

For Dewey it is not only possible to separate religious experiences from established religious institutions; he suggests more strongly that it is our duty to separate the religious impulse from traditional religions, as such religions hinder or "prevent genuine religious experiences from coming to consciousness."[19] Dewey, in striving to separate the religious from religions, is thus creating space for a religious way of being in a naturalistic world. Dewey's proposal should be considered an important forerunner to the kind(s) of religious naturalism that will be discussed and analyzed.

Demarcating Religious Naturalism from Related Approaches

There are several contemporary approaches in the science-religion dialogue, and in debates regarding ecology, that share many important similarities with religious naturalism. Here I will highlight some of the points of convergence, but also some of the significant differences between the nature and aims of religious naturalism and some other current approaches.

Given the nature-centeredness of many religious naturalists, I would like to point out the potential similarities between this religiously orientated naturalism and nature religions. The term "nature religion" designates the overall idea that nature is sacred and worthy of reverent care. This idea is central "to the identities of a number of groups whose participants consider themselves to be engaged in what they also sometimes call nature religion."[20] These religions and spiritualities include paganism, indigenous traditions, new religious movements, and New Age spirituality. The revival of nature religions and pagan traditions is linked to an increase in environmental awareness: "indeed, those who consider themselves to be pagan have been deeply involved in radical environmentalism, including participation in Earth First! (from 1980) and the Earth Liberation Front (from the early 1990s)."[21] Similar to many religious naturalists, people involved in nature religions or pagan/neopagan spiritualities suggest that an eco-ethical system, or practical involvement in the environmentalist movement, flows from having nature as a focus of religious orientation.

However, contrary to religious naturalists, the beliefs of people adhering to nature religions refer often to gods, supernatural beings, and animist ideas, whereby animals and plants have their own spirit and perhaps are constitutive parts of the Divine Spirit. In this way, religious naturalism and nature religions subscribe to radically different ontologies despite both being characterized by a commitment to increasing ecological awareness through religious practices. It should also be pointed out that the religious naturalists in this book, unlike thinkers involved in the variety of neopagan movements, have as their primary aim to develop a religious alternative consistent with science, and in particular a naturalistic understanding of modern science. Religious naturalism occupies more clearly a position in the science-religion dialogue, with its emphasis on harmonizing a religious outlook on life with the major theories of science.

There have been multiple attempts in the dialogue between science and religion to naturalistically conceive, reconstruct, and reinterpret religious ontologies and discourse. Some of these naturalistic approaches to religious worldviews express similar views to that of religious naturalism, as they are committed to reconceptualizing religion in order to avoid potential conflicts with the natural sciences. Drees makes a helpful distinction between "naturalistic theism" and "theistic naturalism," which will be explored below. Before doing so, however, let me briefly comment on what is usually meant by "theism" in these discussions.

Generally speaking, the concept of "theism" intends the view that there is some Supreme Being, mind, or reality that exists beyond the physical universe. This reality is often referred to as "God" and is depicted as the originator of everything that is. However, God did not just create the universe but is also actively sustaining it.[22] In this way, the telos of the natural world, including humanity, is intimately linked to the intentions of God and God's purposes. In the science-religion dialogue we find those naturalists who maintain that such a theistic view is compatible with naturalism, and those who reject such a compatibility-thesis.

Naturalistic theists are those theists committed to naturalism *within* the world but who maintain that physical reality has its origin in a transcendent God. According to this approach, God created the universe and all of physical reality, with its laws, regularities, and properties. Moreover, the physical world is fully describable by the natural sciences, and there are therefore no explanatory gaps within the natural order. The sciences can thus explain all phenomena and events within this world, but not the world as such, as there are limit-questions regarding the origin of physical reality that remain unanswered by the natural sciences. Drees identifies himself as a naturalistic theist. I would suggest that a naturalistic theist accepts local naturalism—that is, one pertaining to this world—but retains a theistic commitment and so rejects global naturalism (regarding reality as a whole).

Should Drees, then, be classified as a naturalistic theist instead of a religious naturalist? It is interesting that out of the three approaches—theistic naturalism, naturalistic theism, and religious naturalism—Drees identifies his approach with naturalistic theism. Niels Henrik Gregersen has also classified Drees's religious naturalism as a form of theism. On this view, "God is so unique and different from the world of creation that no temporal, personal, or causal characteristics can be attributed to God."[23] According to this "atemporal theism," the causal closure of physical reality excludes the possibility of God acting within physical reality, but God may exist beyond the temporal realm.

Gregersen therefore interprets Drees's approach theistically, as Drees leaves open the possibility of God existing beyond the natural order as the originator of the physical universe. Nevertheless, given that Drees's evolutionary account of the origin and function of religion rejects religious realism, one must question how theistic this form of naturalism really is. There is no separate ontological being/reality such as God, according to this approach. Drees, like other religious naturalists (and quite similar to certain American pragmatists), proposes a functionalist account of religion and rejects ontological theism. He should therefore be considered a religious naturalist, which Gregersen refers to as "Flat Religious Naturalism."[24]

The second comparable approach is theistic naturalism. This form of theism is, according to Drees, less dualistic than naturalistic theism. A theistic naturalist, generally speaking, maintains that God is the ground of being and that it is not possible to completely separate the physical from God. This view is strongly related to panentheism, being expressed by prominent thinkers such as Arthur Peacocke and Philip Clayton.

It seems as if the differences between religious naturalism and the more explicit theistic versions of naturalism regard the ontological status of God, the primacy of physical reality, and the relationship between physical reality and God.

A religious naturalist typically rejects the positive ontological status of God and affirms the primacy of physical reality. However, there are those theists who uphold the realness of God while accepting the primacy of the physical. The evolutionary and emergent theist Samuel Alexander (1859–1938) maintains such a view.[25] For Alexander, God or the qualities of a deity emerge through the process of evolution. In this way, emergence theory applies not just to nature but to God as well. Hence, God as an emergent reality is subjected to the laws of nature, just like any other object in the universe. God, as understood on this view, is both transcendent and ontologically dependent on the physical universe. This is an explicit theistic approach, but it shares with religious naturalism the idea of the primacy of physical reality.

What about panentheism or what Drees calls theistic naturalism? Let us first look at emergent panentheism, most notably expressed by Peacocke and Clayton. For these thinkers, God has brought the universe into existence. Yet the universe, or the totality of the physical, is not ontologically separate from God but located "in" God's being. The world is within God and is often construed as God's "body." A panentheist of this type seeks to advance an immanent conception of God. An emergent panentheist suggests, quite similar to a religious naturalist, that supernaturalism is no longer tenable, and that we must therefore find new ways of conceptualizing God and God's providential activity. Still, both Peacocke and Clayton affirm a stronger notion of transcendence, compared to religious naturalists, whereby God is more than the spatial-temporal realm. Hence, this approach affirms both the ontological reality of God and the intimate relation between creation and Creator, while rejecting the idea of physical reality as being metaphysically ultimate.

Process panentheism takes a slightly different approach with regard to the relationship between God and the physical, and this might bring it closer to religious naturalism. Similar to religious naturalism, and some other versions of panentheism, process theologians reject the idea of supernaturalistic interventions, that is, God interrupting, suspending, or breaking natural laws. Unlike emergent panentheists, they reject the idea of creation *ex nihilo*, and the idea of God bringing the universe into existence out of *absolute* nothingness. Process panentheists, such as David Ray Griffin, maintain instead that God created the universe out of "relative" nothingness. This means that our "world, rather than being created out of an absolute absence of finite entities, was created out of a chaotic situation in which there were no 'things' as we ordinarily understand the term."[26] Hence, the "creation of our world, in the first stages, involved the formation of very elementary enduring things, perhaps quarks, out of this chaos of events."[27] It is because of the rejection of dualism between God and the created that process theism, according to Griffin, is significantly less dualistic than many of its religious and metaphysical rivals.

Stone notes that process panentheism shares many similarities with religious naturalism, and that these two approaches might both be described as theologies of immanence. I would conclude with Stone that there are significant differences between religious naturalists and panentheists, be they process panentheists or otherwise. Stone writes, "However, at the end of the day for them [process theologians], there is a 'dimension,' which we humans can call God, that is in some sense not reducible to this world. This dimension does not appear in the group of thinkers that I distinguish as naturalists with a religious bent."[28] That is, a process panentheist, even though subscribing to some form of naturalism, maintains the positive ontological status of God, an irreducible spiritual reality. Although this reality is intimately connected with the physical domain, it still transcends it, and this sets panentheism apart from those naturalists "with a religious bent" discussed in this book.

The Nature of Naturalism

As the term "religious naturalism" suggests, it is a type of naturalism. Naturalism is a philosophical position according to which nature is the ultimate reality. There is nothing beyond nature, and all supernatural or extranatural entities are ontologically excluded from reality.[29] Although naturalism is a philosophical position, it relies heavily on the discoveries and presuppositions of science. It should be pointed out that naturalism is a slippery concept and has been construed in multiple ways within religious naturalism with respect to both ontology and epistemology.

Some, indeed most, religious naturalists have argued for what is called "nonreductive naturalism." This form of naturalism, roughly speaking, affirms the ultimate reality of nature; that the physical world is a "unity in the sense that all entities are made up of the same constituents."[30] This perspective also stresses the irreducibility of what are called "emergent properties," such as consciousness, human culture, the biosphere, normative concepts, morality, and so on.[31] Emergent properties are dependent on the lower levels of reality (for example, consciousness cannot exist without a body). Yet an emergent property "cannot be understood by close scrutiny of the lower abilities."[32] By saying that something is an emergent entity or property, one is claiming that it is not possible to reduce it to the ontology or vocabulary of, for example, physics. Reality is in a sense layered, and so nonreductive naturalism emphasizes the realness and integrity of all these layers.

Contrary to the emergentist position, some proponents of religious naturalism, such as Charley Hardwick and Willem Drees, have argued for a materialist or physicalist interpretation that depicts nature as the ultimate reality. This view also states that every event has a necessary and sufficient physical cause. Hence, determinism is true and every higher level of reality is in principle ontologically reducible to the lower levels. Therefore, our sense of "I" and the belief that we are free, autonomous creatures are biological illusions or byproducts of certain biological mechanisms. According to this view we must reject all dualistic distinctions between mind and matter, and this metaphysical move takes us closer to a stronger form of reductionism.[33]

Despite this rejection of ontological dualism with respect to consciousness and mind, most advocates of religious naturalism with materialist and physicalist inclinations are unwilling to adopt *eliminativism*, which is the view that mental language can be reduced to and replaced by the language of neuroscience. Instead, some proponents of this position argue that we can "save" mental life or mental language by making a distinction between explaining and explaining *away* a phenomenon; that is to say, eliminating it.[34] If this is correct, then we would be able to propose a naturalistic explanation of consciousness without eliminating the mental domain.

Another issue concerning religious naturalism is that of the possibility of proving or disproving the naturalistic position. Is it possible to conclude the truth of naturalism based on the success and deliverances of science?

Willem Drees is among those who answer in the affirmative. He argues that given new discoveries made in neuroscience, all dualistic conceptions of human nature are untenable. William A. Rottschaefer describes Willem Drees's proposed naturalism as a *discovered* naturalism, meaning that we are able to discover the correctness or truth of the naturalistic position by understanding the world as mediated by science.[35]

Jerome Stone, on the contrary, believes that naturalism is unprovable, yet it makes more sense than any other competing worldview, such as theism or dualism.[36] Therefore, it is justified to presuppose naturalism in the dialogue between science and religion. This view of naturalism is quite similar to that of Michael C. Rea, who has defended a view of naturalism *as a research program*. A research program is a collection or set of metaphysical and methodological dispositions that we employ when pursuing certain questions.[37] A research program is not something that we come to adopt on the basis of evidence at the end of our inquiry.[38] Rather, we adopt a research program in order to allow certain questions to be pursued in the first place, to make a certain inquiry possible. It is the research program that decides what counts as genuine evidence. One could, in the light of Rea's definition, argue that Stone treats naturalism as a research program rather than a metaphysical position that we can prove by using the methods of science.

The Epistemological Authority of Science

The question outlined above of whether it is possible to deduce the truth of naturalism from the discoveries of science is connected to the larger issue of the epistemological authority of science. Should we consider the scope and competence of science to be limitless? Or should we be willing to admit that there are areas of reality that lie outside the reach of science? Proponents of religious naturalism seem to differ on this point.

Taking a more reductive stance, Charley Hardwick seems to suggest that there is no limit to the competence of science. He argues that every honest form of naturalism must be based on physicalism, which is itself conditioned by the discoveries of physics. Hardwick criticizes some of his fellow religious naturalists for not taking the advancements of science seriously enough. He complains that many "so-called religious naturalists . . . nostalgically try to smuggle something from the theist tradition into the metaphysical basis for their religious affirmations."[39] For Hardwick, the honest naturalist should willingly concede that "God is nowhere to be found in a physicalist inventory of what exists."[40] Physicalism, according to Hardwick (who is inspired by the philosopher John Post), consists of at least three principles: everything is physical; there is no difference without a physical difference; and all truth is determined by physical truth.[41] It seems that he advocates a *monistic epistemology*, according to which science alone, and especially modern physics, defines the ontology of reality; only science can claim to know what exists and what should be viewed as ontologically improbable or unacceptable.

However, the majority of religious naturalists hold to a pluralistic epistemology, and by that I mean an epistemology that recognizes many paths to knowing reality. Such pluralism urges us to supplement the natural sciences with the humanities and the arts. Stuart Kauffman notes that we have two competing cultures in our society, manifested in the "natural sciences" and the "humanities." He further observes that the latter has come to be recognized as a second-class citizen with respect to our pursuit of knowledge. Kauffman writes, "I believe it is important that this view is wrong. Science itself is more limited by the un-prestatable, unpredictable creativity in the universe than we have realized, and, in any case, science is not the only path to knowledge and understanding."[42] Moreover, concerning the humanities, he writes, "Truth abides here, too."[43] Kauffman is clearly proposing a pluralistic epistemology, and his comments are representative of many religious naturalists.

However, there are some who occupy a middle ground concerning the potential limits of science. Willem Drees argues that we should concede that the arms of science cannot reach every aspect of reality. Thus there are certain "limit-questions," for example, "Why does something exist rather than nothing?" Yet, in a context in which knowledge is attainable, we should go with what the natural sciences say. Science should always be recognized as a privileged source of knowledge, if knowledge is indeed possible.

Religious Aspects of Reality

Above I have outlined a variety of naturalistic positions that are proposed and defended by religious naturalists. I have also described different ways that these naturalists construe the epistemological authority of science. But this is not enough to pinpoint what is religious about this perspective, and, as Michael Cavanaugh correctly notes, "one can affirm naturalism without affirming *religious* naturalism."[44] It is possible to identify at least four ways of affirming religious aspects of reality from the perspective of naturalism:

1. What is religious about nature refers to a specific aspect of it.

2. What is religious about reality refers to nature as a whole.

3. The possibility of a religious conception lies in the limits and explanatory gaps of science.

4. To talk about religious aspects of reality refers to events, norms, or values that are transcendent yet fully consistent with naturalism.

Proponents of (1) argue that we can pick out a certain aspect of the natural order that we may recognize and celebrate as religiously significant. Gordon D. Kaufman represents this position when he argues that we must, given the developments in science and the ecological crisis, change our conception of God as the Creator to what he calls "serendipitous creativity." Kaufman writes, "If God is understood as the creativity manifest throughout the cosmos—instead of as a kind of cosmic person—and we humans are understood as deeply embedded in, and basically sustained by, this creative activity in and through the web of life on planet Earth, we will be strongly encouraged to develop attitudes and to participate in activities that fit properly into this web of living creativity, all members of which are neighbors that we should love and respect."[45] According to Kaufman, God is to be identified with the cosmic and biological evolutionary process, which gives rise to new ontological possibilities.

Karl E. Peters follows Kaufman in understanding God as creativity, as the process of creation.[46] Like Kaufman, Peters suggests that this conception of God may enable us to become more closely related to God by participating "intentionally and completely in the creative process."[47] By participating in this process more fully we are "dancing with the Sacred." God is in this way identified with the evolutionary process itself, and by understanding the structure of evolutionary theory we are able to gain a greater appreciation of nature and of our place in the web of life.

Others, however, seem to consider option (2), whereby nature as a whole is worthy of devotion, love, and respect. Donald A. Crosby suggests that we should "grant to nature the kind of reverence, awe, love, and devotion we in the West have formerly reserved for God."[48] We are, as natural creatures, totally dependent on nature.[49] One could say that nature is of ultimate concern for us; nature is our home.[50] Crosby argues that the whole of nature can function as a religious object for us in the sense that it is appropriate to attribute to nature six basic "role-functional" categories. An object must fulfill these "role-functional" categories, according to Crosby, in order to qualify as religiously significant.[51] These are:

1. Uniqueness, meaning that it lacks a counterpart or is extraordinary.

2. Primacy, meaning that it is of ultimate concern.

3. Pervasiveness, signifying that it integrates every aspect of human life into a whole.

4. Rightness, meaning that it can serve as a standard of critique.

5. Permanence, suggesting that it provides a source for dealing with the insecurities and problems of temporal existence.

6. Hiddenness, which claims that it is in many ways beyond speech and understanding.

On Crosby's vision, nature in its entirety fulfills all these categories and can therefore be conceptualized as an object worthy of religious affection. Moreover, nature is of ultimate concern for us given that everything that is good about humans, such as language, art, science, morality, religion, technology, and philosophy, is made possible by the creativity of nature.[52] Nature, of course, is not morally unambiguous in the sense that everything about nature is good; many evils are evident in the natural world. In response to such concerns, Crosby suggests that these evils "do not diminish or detract from the religious rightness of nature."[53] Indeed, Crosby argues that if nature is to be considered a religious object, it must be, in some sense, ambiguous, since nature's goods and evils necessarily go together.[54] A religious object would be too small if we identified it only with one aspect of reality (namely, what is good in reality).

Option (3) motivates a religious conception of reality based on the explanatory inadequacy of science. That is to say, science has certain limits, and by conceding this we invite the possibility of giving reality a religious interpretation. Willem B. Drees seems to lean toward this view. As was noted above, he argues that science encounters certain "limit-questions" that concern the origin and nature of the world as a whole.[55] Given these epistemological limitations of science, it is justified to postulate "God" as an answer to these kinds of deep metaphysical questions. As long as we formulate a model of divine causality that is not dependent on a supernaturalistic interpretation of God, we are able to maintain a naturalistic worldview. Drees points out that the phenomenon of limit-questions should not be interpreted as evidence in support of a theistic or religious worldview; "however, they could serve as proposals for answers to those limit-questions."[56]

Stuart Kauffman puts a similar emphasis on the boundaries of science. He describes the epistemological problem of predicting the course and direction of the biosphere, the evolutionary process. Kauffman argues that there are too many variables to take into account when trying, for example, to simulate the outcome of the evolutionary process. Kauffman writes that one would have to "carry out infinitely many . . . simulations in order to model our specific biosphere with perfect precision. Obviously, no one could get this much time on a supercomputer."[57] The biosphere and human culture are creative in ways that are essentially unpredictable.[58] This epistemological problem severely undermines the reductionist worldview espoused by, among others, Galileo, Newton, and Laplace. When the reductionist worldview crumbles, the reality of meaning, agency, ethics, and purpose becomes a possibility. Thus we have all the essential elements for a religious worldview and a new conception of the Sacred. There is a strong connection, for Kauffman, between the limitations of science and the possibility of formulating a religious worldview.

Finally, option (4) incorporates religious aspects into naturalism by viewing specific events, norms, or values as religiously significant. As mentioned earlier, Hardwick maintains that all attempts to do theology must be firmly based on a physicalist understanding of reality and ourselves. Such a physicalist way of defining the ontology of reality, with its emphasis on the primacy of physics, carries, of course, certain implications for our understanding of religious language. Theological statements about God can no longer be understood as statements about an objective reality/being called "God." Instead, Hardwick argues that theological statements should be considered "valuational," in the sense that they are metaexpressions "for a form of life that is expressed as a theistic seeing-as."[59] By adopting a theistic seeing-as, one can encounter "events of grace" that offer the possibility of self-transformation, which is an attitude of fundamental openness toward the future.[60]

Karl Peters, in some of his work, seems to adopt a similar position to Hardwick. He says that, in order to understand what spiritual transformation entails, it might be helpful to focus on transformative events in our lives, events that invite a new self-understanding. Like Hardwick (without explicitly connecting this idea to him), Peters calls these events and experiences "events of grace."[61]

Religious Naturalism and Traditional Religion

All religious naturalists seek to formulate a religious alternative that is believed to be more scientifically acceptable than traditional expressions of religion. Most also desire a religious alternative that can motivate humanity to act in more ecologically sensitive ways. It is possible to divide their critique of traditional religion into two kinds. The first critique concerns the ontological and epistemological problems of maintaining a traditional view of God, transcendence, and morality in the light of science. The second critique concerns the ecological inadequacy of traditional religion, which should be interpreted as a pragmatic objection to the great religions as they are presently conceived.

Scientific Objections to Traditional Religion

Proponents of religious naturalism suggest that traditional (supernaturalistic) religion is not scientifically tenable; it is not an option in an age in which science is conceptualized as a privileged source of knowledge, and they have provided many arguments to demonstrate why one should share their view.

Willem Drees maintains that the scientific worldview has shown ontological dualism to be false. He believes that we should take science seriously, and that this in turn should lead to significant changes for religion. He seems to give three different arguments in support of this conclusion.

The first argument is formulated with regard to the problem of finding an adequate model for divine action. Can we coherently claim that God interacts with humanity and the universe when science, as many have argued, seems to reveal a view of the world that is characterized by a "tightly knit web of processes described by laws"?[62] Some have argued (and Drees takes the physicist-theologian John Polkinghorne as an example) that when we consider the causal openness in the universe, as suggested by quantum physics, another conception of reality, in which human and divine action is possible, is beginning to take shape. Drees, however, objects, and argues that the perceived openness in the universe might spring from the fact that we have not been able to explain these "random occurrences" yet but may do so in the future.

According to Drees's second argument, realism is not applicable to the language of Christian theology. Given the success of science, Drees avers, one is justified in believing that science is discovering facts about an independent world. He describes this as the commonsense view of science.[63] But, when one turns to the domain of theology, it is not quite as easy to discern any real progress; thinkers are still having huge disagreements concerning the nature and existence of God, divine agency, how one should respond to evil in the world, the meaning of life . . . and the list goes on. It seems, according to Drees, that we are not justified in believing that theology describes an independent reality.

The third argument concerns religious experience and the cause of these experiences. Drees argues, given discoveries delivered by contemporary neuroscience, that the relationship between mental and physical states is very intimate. This view challenges dualistic thinking and the idea that a phenomenon can have both physical and mental properties.[64] Hence, neuroscientific discoveries carry some important implications for our understanding of religious experiences. The correlation between physical and mental states, or the reduction of mental states to physical states, implies that it is no longer tenable to interpret religious experiences as experiences of God. Thus naturalistic explanations have eliminated God as a candidate for reference.

Another scientific objection to supernaturalistic religion comes from Gordon D. Kaufman. He maintains that the notion of a creator God suffers from severe problems due to insights gained from evolutionary biology. The notion of a creator God implies a view of a conscious being that brought the world into existence. In this sense we have mind before matter. But according to evolutionary theory, the opposite is true: matter brought mind into existence after billions of years of evolution.[65] Thus if one takes the findings of evolutionary biology seriously, the notion of a creator God, a transcendent conscious being that existed prior to the evolutionary process, becomes problematic.

Charley Hardwick also provides some scientific reasons to doubt traditional religion, and specifically all forms of personalistic theism, which stipulates a transcendent being that has all those traits that we take to be personal. He says that naturalism, and science in general, can account for all that exists without invoking an intelligent mind with intentions and ideas. Hardwick claims, "All existence, all order, and all action can be accounted for without recourse to the operation of intelligent purpose."[66] On this view, theism is explanatorily obsolete given the sufficiency of science.

Pragmatic Objections to Traditional Religion

The second objection that religious naturalists bring up against traditional forms of religion is more pragmatic in nature. Loyal Rue asks whether the world's received traditions have the capacity to give a morally relevant response to the environmental crisis. According to Rue, who cites Lynn White Jr.'s famous 1967 article in the journal *Science*, the Judeo-Christian tradition is uniquely responsible for the environmental crisis because it has promoted the idea that the sole purpose of nature is to serve humans.[67] Rue concludes that, despite so far having neglected ecological responsibility, these traditions do possess the necessary resources to inspire a morally adequate response to the crisis. Rue argues that nevertheless the likelihood of this being realized in practice is not promising given the expansion of conservative religious movements that view the increase of environmental consciousness as a sign of secularization.

Similarly, Karl E. Peters writes that traditional religion has given rise to numerous problems concerning both the environment and the spiritual life of many people. Peters, who writes from the perspective of American pragmatism, believes that developing a new conception of God may help us not only respond to the environmental crisis but also have a more satisfying religious life. The dualistic mindset that has been pervasive in traditional religion, according to Peters, has led many to identify the natural world as a source of temptation, and even as evil.[68] God, as described by traditional religions, has been criticized for seeming like an "absentee landlord" who has left the scene after bringing the universe into existence.[69] God has been depicted as a distant creator, as a designer who created nature for us to conquer and manipulate.[70] This has led to many bad decisions on the part of governments and individuals with respect to the health of the ecosystem. Peters concludes that if religion is going to be a part of the solution, and not the problem, new images of God are required.

Like many in this debate, Gordon Kaufman maintains that much is at stake in a time of ecological decline. He argues that Christian theology has been too human-centered, structured around an anthropocentric conception of God. This human-God relationship, which has been emphasized in theology, has in Kaufman's words, "obscure[d] and dilute[d] . . . ecological ways of thinking about our human place in the world."[71] We need, instead, a new conception of God that can help us to highlight our dependency on nature and the ecosystem. As a consequence of this destructive theology, many have tended to view nature as a rival to a devotional relationship to God, and, according to Kaufman, this view must be abandoned.

Global Religious Naturalism?

In light of the scientific and pragmatic objections that have been raised against traditional expressions of religion by these spiritually inclined naturalists, one should wonder if advocates of religious naturalism propose a complete rejection of traditional religion, or if they recognize some force and validity to it. Some have asked, "Should religious naturalists advocate a naturalistic religion?"[72] It is possible to identify three different answers to this question:

1. Religious naturalism should be incorporated into an already existing religious tradition (naturalistically informed religion).

2. Religious naturalism should adopt symbols, concepts, and images from other religious traditions (religiously informed naturalism).

3. Religious naturalism should reject the great religious traditions entirely in favor of a globalized form of religious naturalism (global religious naturalism).

Naturalistically Informed Religion

Those who hold to the view of naturalistically informed religion maintain that the future of traditional religion depends on it reinterpreting itself within the framework of naturalism. Gordon Kaufman can be counted as one of those who argue that traditional religions should be reformed on the basis of a naturalistic ontology. He seeks to transform the Christian tradition from within, and maintains that the Christian concept of God, as traditionally defined, must change given discoveries in evolutionary biology, but also in light of the environmental crisis. It is not necessary to reject the Christian tradition completely; if it is modified (or naturalized) successfully it can inspire people to be more ecologically responsible. Kaufman argues that we have to give up the anthropocentric view of God in favor of a postmodern view of God as creativity, a metaphor for the cosmological and evolutionary process.[73]

Some proponents of this view have focused on certain ideas of salvation in Christian theology and considered how to reconcile them with modern science. For Karl Peters, the search for spiritual wholeness and reconciliation with God is the primary concern of religious communities, and in particular the Christian faith.[74] Peters suggests that we can gain a better understanding of salvation by relating it to the scientific view of humans as social creatures. Salvation, the saving transformation, is redefined as moving "from poorer functioning in a less harmonious manner to more harmonious and mutually enhancing relationships."[75] Living in such a harmonious relationship includes not only humans but nature and God as well. Why, then, should we take salvation to be necessary for humans? According to Peters, Darwin's theory of evolution demonstrates clearly that social animals are not able to survive apart from some kind of supportive community. In order to survive we have to connect to a community that is greater than ourselves, that is supportive of caring relationships. This is where the Bible can play a significant role by providing the Christian community with "the criteria of cooperating, caring, and loving universally and unconditionally."[76]

While Kaufman and Peters rely on an emergentist and nonreductionist version of naturalism in order to synthesize the Christian tradition and naturalism, others opt for a more reductive framework. Hardwick's project is an attempt to reformulate Christianity through a purely physicalist ontology. This, it should be said, is a more minimalistic approach to religious naturalism. Hardwick's main ambition is to integrate a physicalist ontology within the Christian tradition so as to make it compatible with the framework of science.

Religiously Informed Naturalism

The perspective of religiously informed naturalism, which is proposed by both Stuart Kauffman and Willem Drees, suggests that religious naturalism can be supplemented and enriched by traditional religions.

The core argument of Stuart Kauffman is that we urgently need to reinvent the notion of the Sacred or God to mean the Creativity of the Universe. Kauffman notes that there is a connection between what we call a thing and how we treat it. By viewing nature as Sacred or equating it with God we will inspire people to act in an ecologically responsible manner toward nature since "no other human symbol carries the power of the symbol, God."[77] Moreover, we must build a shared global ethic, which we have lacked for too long, but which has become a real possibility due to the advances made in science, a deeper understanding of our evolutionary past, and our place in the natural order. Such an ethic must, according to Kauffman, embrace different cultures and religious traditions in order to be recognized as valid. Thus Kauffman suggests that cultures and religions should supplement religious naturalism. Some beliefs in religious traditions are, however, not compatible with this pursuit for a global ethic. For example, the belief that God created all nonhuman creatures for human benefit, that nature is created for us to use and manipulate, must be rejected. We should, on the contrary, concede that "we are of the world, it is not of us."[78]

For Drees, not only is it feasible for a religious naturalist to build on various religious traditions but "such a pluralistic move is actually preferable."[79] However, Drees is also open to the possibility that religious naturalism might develop into a "globally shared system of ethical beliefs," and even into "everybody's story."[80] Nevertheless, this does not have to amount to a negative relationship with other religions, or a dismissive attitude toward the wisdom embedded in such religions. Rather, argues Drees, "whatever is needed is drawn eclectically from the wisdom of the spiritual traditions taken together as a huge reservoir."[81] Drees's view seems close to global religious naturalism, though in the end his desire to eclectically draw on the belief systems of world religions, rather than making naturalism into its own myth, places him within the category of "religiously informed naturalism."

Whereas Kaufman and Peters started with traditional religion and sought to reform it from within by incorporating the central tenets of naturalism, Kauffman and Drees begin with naturalism and seek to add on religious components that they believe will be of benefit.

Global Religious Naturalism

The third approach is to reject all religious traditions as they are presently conceived in favor of a globalized form of religious naturalism. This kind of naturalism would produce its own myths, images, and symbols independently of existing religious traditions. This view is fairly uncommon, but some remarks made by Loyal Rue, Donald Crosby, and Ursula Goodenough can be interpreted as an endorsement of such an approach.

Rue maintains that the ecological challenges are globally relevant and that they require a global solution. He therefore asks if the world's received traditions could provide encouragement when it comes to developing a morally relevant response to the crisis. He is quite skeptical concerning this possibility and claims that there is "little hope that they will make a meaningful difference."[82] The skepticism that Rue expresses is due to the fact that there is no unified theory of nature to be found among the religious traditions.

Rue finds it hard to specify what traditional religions can contribute in an age of ecological decline. In response to those who argue that religion can enhance naturalism (religiously informed naturalism), Rue writes, "But if the question is what Christianity can offer to a religious naturalism that is keenly mindful of the radical urgency of our global environmental crisis, we might consider looking elsewhere."[83] Instead we should identify the core of this posttraditional religiosity, which is the view that "Nature is the sacred object of humanity's ultimate concern."[84] This could be interpreted as a view according to which religious naturalism ought to be considered independently of traditional religions, and that it may one day become a robust mythic tradition.

Nature, as we have seen, is interpreted by Crosby as humanity's religious object. In claiming that reverence for nature is the core feature of religious naturalism, and suggesting that a "religion of nature" can exist independently of the major world religions, Crosby seems to opt for a version of global religious naturalism. Crosby is less keen on dialoguing with already existing traditions, but seeks instead to formulate a naturalistic religious alternative that is on par with traditional religions. Nature *itself* offers meaning and values, and according to this approach there is no need to incorporate religious images or doctrines from other religions.[85]

Ursula Goodenough concedes that traditional religions, in particular the old wisdom traditions, still carry transformational power. Religious metaphors can still resonate with our religious selves, but such metaphors must not be taken literally. To take religious metaphors literally is to adopt the dangerous mindset of religious fundamentalism, according to Goodenough.[86] She further recognizes that the ecological crisis, being a global problem, requires a global solution. Goodenough's ambition to uncover the sacred depths of nature involves developing a global version of religious naturalism that, in a similar way to Rue's and Crosby's proposals, can exist independently of religious traditions.

Rue, Crosby, and Goodenough all seem to advocate for a global version of religious naturalism, which can exist independently of religious traditions. Global religious naturalism is therefore more optimistic regarding the religious power and potential of naturalism. This chapter now moves on to consider the various ways that religious naturalists construe the function of religious language.

The Function of Religious Language

As we have seen so far in this chapter, religious naturalists maintain that religion has to be modified according to a naturalistic framework in order to be scientifically viable. More specifically, they claim that religion needs to overcome its supernaturalistic heritage. As we have seen in the previous section regarding traditional religions, some, but not all, also wish to use or modify religion in order to respond to, cope with, and even heal the current ecological crisis. As a result, we should not take "the Sacred" to refer to some supernatural reality or being, existing independently over and against nature. Instead we should view God, the Holy, or the Sacred as part of the natural order.

This turn from supernaturalism to naturalism seems to have implications for how we should view the function of religious language. Religious naturalists commonly provide two reasons for their new conceptualization of religious language. The first concerns the problem of maintaining the positive ontological status of a supernatural reality and/or being due to scientific discoveries. The second concerns the need to find adequate conceptions of God or the Sacred in order to respond properly to the ecological challenges.

I will argue that some religious naturalists seem to hold to the view that the primary function of religious language is not to describe reality or a part of reality that we take to be religiously significant. This departs from traditional realism and tends toward something more pragmatic in nature. Religious naturalists offer us a new view of religious language. By their proposal, the function of religious language is to invoke an attitude toward nature that is ecologically adequate. I will call this an ecologically mindful attitude (EMA).

The sections below have two main goals. First, I will outline more specifically why religious naturalists seek to develop new understandings of religious language. Second, I will explain the pragmatic approach to religious language, which involves developing EMAs. The development of EMAs is not paramount to all religious naturalists. However, for a significant number of naturalists, this idea defines their approach to religious language.

Scientific Reasons for a New View of Religious Language

In this section, I will briefly describe the way three religious naturalists modify the use of religious language on account of their commitment to metaphysical naturalism.

As mentioned above, Charley Hardwick claims that "God is nowhere to be found in a physicalist inventory of what exists."[87] God pictured as a supernatural being cannot be located anywhere in reality; supernaturalism has been found wanting, and science does not need to invoke any personal explanations in order to account for parts of reality, or even reality as a whole. Most religious naturalists, however, say that we can ground God or the Sacred in nature, which enables one to still affirm the central tenets of naturalism while holding to a religious conception of reality. In contrast, Hardwick is hesitant to adopt this view, since he finds "nothing referentially significant in a religious sense about nature as a whole or nature in its parts."[88] Instead he opts for "valuational theism" in which religious propositions such as "God exists" are expressions of a certain self-understanding. For Hardwick, "God exists" expresses an understanding of us in relation to the rest of the world. Accordingly, "God exists" and similar religious utterances carry no metaphysical implications.

Contrary to Hardwick, Gordon Kaufman and Stuart Kauffman maintain positively that nature is a good candidate for being viewed as a religious object. To them nature is worthy of religious reverence, love, and devotion. As outlined previously, they offer some scientific reasons for replacing the traditional concept of God as a supernatural being with a naturalistic conception of God. This natural "God" would instead refer to cosmological and biological evolution. According to Kaufman and Kauffman, religious language still refers to some aspect of reality, even though all notions of possible supernatural reference must be rejected.

To conclude, several proponents of religious naturalism suggest that we have to find new ways of thinking about language in order to develop a religious alternative that can be considered a live option in an age in which science sets the boundaries for what is believable and acceptable.

Ecological Reasons for a New View of Religious Language

The ecological imperative is the second reason many religious naturalists want to develop a more pragmatically informed view of religious language. Karl Peters maintains that we should find new ways of talking about God and the Sacred so that we may further contribute to the well-being of the ecosystem. He recognizes that there is a connection between how we talk about nature and how we treat it, and by ascribing to nature the same status as we have previously ascribed to God, it will be easier to promote a responsible approach to nature that takes the ecological challenges seriously. Peters therefore highlights the moral implications of religious language.

Gordon Kaufman, who gave some scientific reasons for doubting a traditional view of religious language, also provides some ecological reasons for consideration. His main contention is that we need to include the emerging biohistorical understanding of human life in order to meet several global challenges. To think that "God" refers to a supernatural being has only made us ignorant of ecological ways of thinking of our place in nature and reality. But this can be corrected if we instead come to view God as serendipitous creativity, as the biological processes that led to the emergence of life itself. It is noteworthy that for Kaufman a scientifically informed naturalism goes hand in hand with his ecological concerns.

The message provided by several religious naturalists, therefore, is that the ecological challenge provides certain restrictions when it comes to constructing conceptions of God or the Sacred. Any attempt to reconceptualize traditional religious beliefs, the nature of God, and God's relationship to nature and humanity must consider our ecological situation. Thus we should judge religious conceptions on the basis of how well they support humanity in addressing these pressing global issues.

Ecological Mindfulness and Religious Language

Religious naturalists believe that both science and ecology place restrictions on theological constructions. Consequently, religious naturalism is bringing forth a new way of understanding religious language.

On this view, what is important in the realm of religion and spirituality is to adopt an appropriate attitude to the ecological crisis. Loyal Rue proposes that we adopt an eco-centric morality that views the integrity of natural systems as having absolute value.[89] Similarly, Karl Peters suggests that we have to become responsible "social-ecological selves" and recognize our place in reality and our relationship to our natural family. This is so that we may further contribute to the well-being of human culture and the ecosystems of the earth.[90] Ursula Goodenough argues in a similar way for a planetary ethic.[91] By being mindful of the findings in biology, such as the fragility of life and its ecosystems, the interconnectedness of all life, the need for personal wholeness and social coherence, and so on, we can develop a religious alternative that is up to the task in an age of ecological decline.[92] We need to adopt an attitude toward nature that is ecologically sound; an ecologically mindful attitude (EMA).

By placing the religious aspects of reality in nature (by speaking of nature as divine or sacred), and by avoiding religious language that has been based on dualistic theological constructions, several leading religious naturalists argue that we are more likely to adopt an attitude that serves the stability of the ecosystem. They seem to propose a practical or pragmatic theory of religious language. That is to say, the function of religious language is not only (or maybe not primarily) to describe those aspects of reality that we find religiously significant but rather to "evoke an appropriate dispositional attitude" toward nature.[93] To adopt an EMA is to have such an attitude.

How we use language affects our thinking; language shapes our mind. Language plays an important part in the construction of what Karl Peters calls our "cultural selves," which includes ways of growing, processing, and being religious.[94] The underlying logic is: how we think, so we will act, and how we speak, so we will think. There is an intimate connection between language and action, between how we speak about nature and how we come to treat it. Hence, a great task for thinkers today is to reconstruct religious language in order to inspire people to adopt attitudes that are ecologically beneficial. This is what Gordon Kaufman calls the constructive task of theology: "Theology now becomes essentially a constructive task, and the symbol-systems of our various religious and secular traditions, in terms of which we do our thinking and acting, our living and our worshipping, have to be reconsidered in light of these problems that so urgently demand our attention."[95] According to this pragmatic theory of religious language, truth or what reality is like should not be the primary focus of our religious or spiritual quest. Instead, we should construct and adopt those religious images and conceptions of reality that will further our pragmatic goal(s), which, for many religious naturalists, is a healthy ecosystem in which biological organisms can live and flourish.

Scientific Realism in Religious Naturalism

In light of the changes that religious naturalists believe must take place with regard to traditional religion, it is worth considering whether one should interpret religious naturalism realistically or antirealistically. By "realism" I mean a view that consists of at least three different theses.[96] *Ontologically*, the realist affirms that there is a reality external to human minds and that it exists independently of human thoughts, concepts, and opinions about it. According to the realist, we discover reality; we do not construct it. *Epistemologically*, the realist holds that we can come to know what reality is like, that knowledge of reality is possible. *Semantically*, the realist affirms the possibility of successfully referring to reality. That is to say, we can make statements about reality that can be either true or false. It is possible to identify two versions of realism in the context of religious naturalism; one concerns science, while the other relates to the realm of religion.

Bas van Fraassen defines scientific realism in the following way: "It characterizes a scientific theory as a story about what there really is, and scientific activity as an enterprise of discovery, as opposed to invention."[97]

This seems to be the default position among religious naturalists. Willem Drees holds to this view of science, which he claims to be the "commonsense view." In his words, this means that "scientists are not making up a story, but they are finding out about a real world."[98] The results of science are not fiction, so to speak. Similarly, Loyal Rue presupposes this view of science in his view of scientific materialism and the idea of reaching a unity of science.[99] Gordon Kaufman seems to affirm scientific realism when he argues that science has the capacity to enlighten our understanding of cosmological and biological evolution. And finally, Charley Hardwick suggests that physicalism requires a realist conception of truth.[100]

Realism, Antirealism, and Pragmatic Realism

The truth of scientific realism seems rather unchallenged among religious naturalists; most, if not all, seem to hold to this view. Instead the disagreements seem to be centered on what can be called "religious realism." This sort of realism refers to the view that: (1) religious utterances are fact asserting, and not merely expressive; (2) the truth of those statements is determined by nonepistemic factors; (3) what is the case is independent of human thoughts and ideas; and (4) we can in principle have true beliefs about a reality that exists independently of human cognition.[101]

A religious antirealist would in this case be someone who denied any of 1 through 4. Both antirealistic and realistic interpretations are well represented among religious naturalists. Furthermore, it seems that one can identify a third position, which occupies a middle ground somewhere between realism and antirealism. We can call this middle position *pragmatic religious realism*, and it will be explained further below. This discussion is not exhaustive and does not seek to discuss all religious naturalists with regard to the issues of realism, antirealism, and pragmatic realism. In chapter 4 I will explain in more detail the religious content of each religious naturalist and how their religious views relate to the realism-versus-antirealism debate.

Who are the religious antirealists? Charley Hardwick seems to endorse an interpretation of religious discourse that does not hold to 1 through 4 as outlined above. He is reluctant to ground religious discourse in reality. According to Hardwick, the term "God" does not refer to anything that actually exists; instead the proposition "God exists" is an expression of a "theistic seeing-as" or "taking-as," meaning that we come to view the world in terms of the Christian story of sin and grace.[102] What Hardwick is suggesting is that the dispute between atheists and theists does not concern the truth or falsity of the proposition "God exists"; rather it is about two conflicting ways of existing or ways of being in the world. Christian faith is in this sense a valuational stance and an orientation toward the source of human good.[103] "God," therefore, refers to an expression of a certain attitude. This means that the normative truth (but not the ontological truth) of "God exists" is the description of the human condition characterized by a "self-defeating order of life."[104] Hardwick's project is thoroughly antimetaphysical in the realm of religion, as religious truth is about values rather than ontology.

In contrast, Loyal Rue argues for theological realism and believes that a nonrealistic interpretation is a mistake that poses a "major threat to the effectiveness of mythic traditions."[105] He believes that we must find ways of reinterpreting religious discourse from the viewpoint of naturalism while avoiding a "creeping non-realism."[106] Traditional religions, such as Judaism and Christianity, are dependent on realism as a presupposition for them to function properly. In the absence of a realistic understanding of their root metaphor, "God," these traditions will not function in the way they are intended to, which is to produce spiritual transformation. Rue's solution to the problem of nonrealism is to understand "Nature" as humanity's sacred object, which he believes is compatible with both naturalism and realism. Consequently, realism is for Rue a necessary presupposition for religious naturalism, if it is to produce the desired religious attitudes and behaviors.

A significant number of these naturalists that I will discuss and analyze seem to propose what is called "pragmatic realism."[107] This view, roughly speaking, is a synthesis of realism and pragmatism. It is pragmatic in the sense that it takes into account our human dispositions, our ways of being in the world, and how our human practices actually work. Part of this means, according to the pragmatist, that we are inevitable realists. As human beings living our lives in and through different practices, we tend to presuppose some form of commonsense realism, that is to say a belief in an external world.[108] However, the pragmatist adds that this does not amount to metaphysical realism, as this kind of realism is merely a commitment that we naturally have and operate from. The emphasis is on *commitment* rather than *ontology*. By placing the emphasis differently, realism becomes subordinated to pragmatism, which means that we can escape the pitfalls of both metaphysical realism (the belief in an unconceptualized reality) and idealism (the view according to which reality is completely constructed by humans).

One thinker who seems to adopt pragmatic realism with respect to religious discourse is Gordon Kaufman. As we have seen, a main contention of Kaufman's is that the traditional realistic conception of God must be abandoned for both theological and scientific reasons. We should instead identify God with the biological creativity in the universe. God is in this way *constructed*, which suggests that Kaufman endorses a type of religious constructivism. Furthermore, Kaufman's theological model of God is related to pragmatism as it is constructed against the background of our pragmatic goals and human needs, which are, among other things, ecological stability and well-being. Moreover, this is still realism, since "God" is referring to an external reality, or aspects of it. But we commit to this view of God and reality because of certain pragmatic interests vital for the function of our human practices. Thus one should conclude that Kaufman is proposing pragmatic realism with respect to religious discourse.

Stuart Kauffman, who follows Gordon Kaufman in identifying God or the Sacred with the physical creativity in the universe, seems to hold to pragmatic realism as well. Kauffman argues that God is a construction. But that does not make God any less real, since "God" properly refers to a real aspect of nature.[109] At the same time, "God" is a *chosen* symbol for the creativity of the universe, and this choice is made due to current ecological challenges. In order to reinvent the Sacred, Kauffman combines pragmatism, realism, and constructivism.

I therefore suggest that it would be mistaken to assume that Kauffman and Kaufman propose an antirealistic understanding of religion and religious discourse. The religious reality is real, although we should understand its realness in pragmatic terms rather than strict metaphysically realistic terms, which has been the dominant position in most traditions of theology and philosophy.

Conclusions

In this chapter we have covered several core issues within religious naturalism. I briefly outlined some historical predecessors of religious naturalism, and viewed it in relation to a few relevant neighboring perspectives.

We have identified three possible approaches concerning the epistemological authority of science as it is conceived in religious naturalism. According to one view, science is limitless (monistic epistemology). Another view says that we should concede that even science has its limits and that some areas of reality fall outside the competence of science (pluralistic epistemology). According to a third alternative, science has certain limits, although we should grant that science is limitless with respect to those areas that the scientist is able to investigate.

Another core issue examined in this chapter relates to the religious dimension of religious naturalism. It has been suggested that religious naturalists affirm the religious dimension of naturalism in four different ways.

The relationship between religious naturalism and traditional religion has been discussed. I divided the critique against traditional religion into two kinds. The first concerns the ontological and epistemological problems of maintaining a traditional conception of God, given certain discoveries made in the sciences. The second concerns the ecological inadequacy of traditional religion, and was considered a pragmatic objection.

Connected to these critiques is the question of whether religious naturalism implies a complete rejection of traditional religion. That is to ask, "Are they proposing a religion of their own?" Three responses to this question have been discussed. According to the first response, traditional religion should incorporate insights from naturalism (naturalistically informed religion). According to the second response, religious naturalism can be supplemented by existing traditions (religiously informed naturalism). The last response suggests that religious naturalism may develop into a robust mythic tradition with symbols and images of its own (global religious naturalism).

Furthermore, this ongoing movement within the science-religion dialogue has certain implications for how we should understand religious language. I have argued that in religious naturalism, the function of religious language is not primarily to describe a religious reality that exists independently of the natural world. Instead, some religious naturalists maintain that the function of religious language is to evoke an appropriate dispositional attitude toward nature. This has been called an "ecologically mindful attitude."

Lastly, we have covered the issue of realism and antirealism. I proposed that we could identify both realistic and antirealistic interpretations of religious discourse. However, a third approach has been discovered that seems to occupy a middle ground between realism and antirealism. This has been called "pragmatic realism," and it seems to attract several religious naturalists. Having outlined the main features of religious naturalism, and central debates between these religiously committed naturalists, I will now move on to a more thorough analysis (chapter 2) and evaluation (chapter 3) of the naturalistic aspect of religious naturalism.

2

What Is Naturalistic about Religious Naturalism?

The issue that will be addressed in this chapter concerns the relationship between naturalism and science as it is conceived and argued for by religious naturalists. As previously discussed, religious naturalists argue that science, and specifically the natural sciences, challenges traditional dualistic and supernaturalistic conceptions of reality. Today, they argue, science tells a story about a world that is empty of supernatural beings and realities. It was suggested in the previous chapter that two forms of naturalism seem to be employed in this discussion: a reductive or physicalist/materialist version of naturalism, and a nonreductive or emergent naturalism. This chapter aims to explicate the arguments behind these forms of naturalism so that we can more fully understand religious naturalism as a position in the science-religion debate. In this chapter, the reductive version will be referred to as *monistic* naturalism, while the nonreductive version will be called *pluralistic* naturalism. However, before outlining the different claims of monistic and pluralistic naturalism, I will give a brief account of the central pillars of the naturalist tradition and three arguments that are proposed by religious naturalists to support the ontology under consideration.

The Central Pillars of Naturalism

Naturalism has a long history and stretches back to Greek philosophy with figures such as Democritus, who maintained an atomic theory of reality, and the Roman poet Lucretius. Naturalism can also be found more generally in Epicurean philosophy. It is not easy to define naturalism, and it is often identified or equated with materialism, empiricism, and sometimes with scientism.

A paradigmatic formulation of naturalistic ontology was given by Ernest Nagel in his lecture "Naturalism Reconsidered." Nagel suggests that there are two central theses for naturalism: the first is the "existential and causal primacy of organized matter in the executive order of nature."[1] This thesis suggests that all human behavior can be analyzed and explained in terms of the organization of temporally located bodies "whose internal and external relations determine and limit the appearance and disappearance of everything that happens."[2] The second thesis is that the manifest plurality of things and their properties and qualities are not illusory but irreducible features of the cosmos. However, such irreducible phenomena cannot be explained in terms of some supernatural reality beyond the physical. They are not "a deceptive appearance of some more homogenous 'ultimate reality' or transempirical substance."[3]

Naturalism, in rejecting transempirical substances and supernaturalism, affirms a general kind of monism, meaning that there is one fully natural reality. Arthur C. Danto writes concerning the core features of naturalism, "Naturalism, in recent usage, is a species of philosophical monism according to which whatever exists or happens is *natural* in the sense of being susceptible to explanation through methods which, although paradigmatically exemplified in the natural sciences, are continuous from domain to domain of objects and events. Hence, naturalism is polemically defined as repudiating the view that there exists or could exist any entities or events which lie, in principle, beyond the scope of scientific explanation."[4] This monism is often coupled with the idea of "the causal closure of the physical," that "the cause of an entity or event in the spacetime world is itself within spacetime world . . . there is no causal intervention from outside the spacetime world."[5] Because of this commitment, divine causal interventions have been ruled out.[6] This is a defining *ontological* commitment for naturalism.[7]

A defining *epistemological* feature of this ontology, and for religious naturalism, is the reliance on science in the area of knowledge. This view was strongly defended by Willard Van Orman Quine, who rejected philosophy as the primary source of knowledge. He wrote that naturalism entails the "recognition that it is within science itself, and not in some prior philosophy, that reality is to be identified and described."[8] Many have tried to define naturalism *negatively*, suggesting that it merely rejects supernaturalism and ontological dualism. Quine, however, adds a positive dimension to naturalism, namely that naturalists affirm a strong belief in the epistemological authority of science. Science is the most reliable guide for uncovering reality, and its reach extends into other disciplines.

The forms of naturalism that I will evaluate do not merely express methodological doctrines; they also make ontological claims about the nature of reality, as well as epistemological claims regarding the epistemic effectiveness and reach of science. Although it might be possible to derive methodological doctrines from monistic and pluralistic naturalism regarding the practice of science, this is not their emphasis. Hence, I will not seek to discuss the validity of methodological naturalism but rather focus on the soundness of these monistic and pluralistic positions with regard to potential metaphysical grounding problems.

Why Should One Be a Naturalist?

Roy Wood Sellars once claimed, in relation to the impressive success of the physical and biological sciences, that we all seem to have become naturalists.[9] Whether Sellars's observation is correct is a matter of debate, but no one today can argue against the fact that the natural sciences have changed our understanding of reality and the place of human beings in the natural order. Religious naturalists would prefer that the science-religion debate move in the direction of naturalism and away from dualism. But *why* should we adopt naturalism? Why should we prefer naturalism over any other competing worldview or metaphysical framework? In the science-religion debate it is possible to identify a number of reasons that have been proposed in favor of the naturalistic position. According to one view we should adopt naturalism because it is an economically effective explanation. A second view states that we should adopt naturalism because that is the view that seems to be emerging out of science. We may not be able to derive the truth of naturalism based on science, but it seems to offer the best framework for dealing with questions that have arisen in various scientific practices. Others argue that naturalism is the best candidate due to its pragmatic advantages. I will present each line of reasoning below.

Naturalism as an Economical Explanation

Naturalism has been described and defended by many as an economical explanation. Some suggest that naturalism, as an explanatory hypothesis, should be accepted since it stays close to our knowledge about the natural world. We should, according to Chet Raymo, "never suppose a complex explanation when a simpler explanation will suffice."[10] This statement is, of course, an expression of Ockham's razor, a principle that suggests that we should shave away any superfluous explanations, theories, entities, and so on in our scientific pursuit. Religious naturalists maintain that we live in a natural world free of gods, ghosts, and spirits, and to claim something else would therefore be considered extraordinary. As Carl Sagan famously wrote, "Extraordinary claims require extraordinary evidence."[11] By this way of reasoning, the naturalistic story is simpler than its supernaturalistic counterpart and should therefore be presumed true until we have positive reasons to suppose something else.

The point of using Ockham's razor is to achieve economy of explanation by shaving away unnecessary concepts, categories, and principles that have not been established by science or that are not a part of the scientific story. Loyal Rue writes, "Thus, naturalists oppose explanations that unnecessarily assume a transcendent order of entities and events having causal influence in the order of nature. Why posit two orders of being where one is sufficient?"[12] Supernaturalism would in this way introduce concepts and categories that are unnecessary with respect to our understanding of the world. We can make sense of reality without invoking entities or explanations that lie outside of the natural order.

Supernaturalistic claims once enjoyed widespread acceptance among cultures, but science, with its commitment to the view of the universe as describable in terms of natural causes, has changed the situation dramatically. Nowadays claims of supernatural occurrences are met with skepticism, and those that make such claims, according to this line of reasoning, have to produce strong evidence to support them.

Thus this argument concludes that naturalism seems to offer the best interpretive framework for dealing with questions concerning the nature and origin of reality, because it refrains from positing unnecessary and unverifiable ontological categories.

Naturalism as Implied by Science

The second argument concedes the fact that naturalism is a metaphysical position, meaning that the truth of naturalism is not strictly deducible from the major theories in science. However, this does not mean that naturalism should be considered a metaphysical framework that one adopts beforehand so as to decide how to interpret the findings of science, or to try to make empirical theories fit the naturalistic story. Rather, as Willem Drees puts it, naturalism is low-level metaphysics "in that it stays close to the insights offered and concepts developed in the sciences."[13]

Naturalism can thus be construed as an empirical hypothesis, which means that the status or success of naturalism is dependent on the development of science. Naturalism could therefore be falsified if science would move in the direction of introducing nonnatural concepts or categories, or if scientists would find reasons to believe that certain phenomena seem to transcend the natural order or that something could not be accounted for in terms of natural causes.

Drees maintains that naturalism is supported by contemporary science, and that it is the metaphysical framework that is most faithful to the practice and discoveries of science. It is for this reason that he argues that naturalism is preferable to other competing ontologies.

Pragmatic Benefits of Naturalism

According to a third possible argument, we should adopt naturalism because it carries certain pragmatic benefits. Philip Clayton recommends that we abandon all forms of supernaturalism and embrace naturalism for the sake of the practice of science. He argues that there is an initial presumption in favor of naturalism. Clayton writes concerning the motivation behind this presumption, "If we do not make it, science as we know it would be impossible. Scientific activity presupposes that causal histories are reconstructible in principle, which they would not be if the cause of some specific phenomenon lay outside the natural order altogether."[14] Pragmatically we ought to be embracing naturalism, given that the practice of science seems to be dependent on a naturalistic perspective of the world.

Willem Drees claims, in a similar way, that we should adopt naturalism for pragmatic reasons. One reason, according to Drees, is that naturalism, by aligning itself with the best available science, has been able to produce impressive technological achievements in recent years.[15] By accepting naturalism, and by virtue of its epistemic commitment to the methods of science, we are taking science seriously. According to Drees, this is something we have to do in order to combat modern catastrophes such as the AIDS epidemic, as well as to find effective solutions to the ecological crisis. There are thus practical reasons to take into account when considering the arguments for naturalism.

While some naturalists emphasize the pragmatic benefits with respect to science and technological achievements, others highlight the pragmatic benefits of naturalism for our religious and spiritual life. Naturalism, they suggest, is not only compatible with a religious outlook on reality but actually has certain advantages over some of its metaphysical rivals, such as theism. Karl Peters argues that traditional theism, because of its dualistic metaphysics, has made it difficult for many to relate to God or to even speak of God. God in the tradition of classical theism has been described as the originator of the Big Bang. However, since the event of creation the God of classical theism has been absent from nature and uninvolved with history. Peters writes, "The source of all existence becomes a source only at the beginning of the universe and not a source of new forms of existence on our planet, in human history, and in our daily lives. God thus becomes like an absentee landlord, or a developer that leaves the scene."[16] According to naturalism, if God exists, then God must be conceived as the process of creation itself, not an ontologically distinct being. There is no longer an ontological gap between humanity and God, which means that we "can become more closely related to God."[17] This way of arguing in the area of religion becomes a pragmatic reason for adopting naturalism.

In this section we have seen two types of pragmatic reasoning. One is concerned with using the advances in science in tackling real-world problems. Moreover, it was suggested that a belief in naturalism is a prerequisite for scientific practice and progress. The other pragmatic argument focused more on religion and spiritual life, and on the religious benefits of adopting a naturalistic conception of God.

Monistic Naturalism

We will now more thoroughly describe and compare how key proponents of religious naturalism express the two main types of naturalism: monistic naturalism and pluralistic naturalism. In so doing, I will reconstruct these positions by collecting different beliefs pertaining to the nature of naturalism. These are then presented as propositional statements regarding ontology, epistemology, and semantics.

An advocate of monistic naturalism typically claims that we should come to view the world from the perspective of science and that we should construct accounts of reality solely in terms of natural facts and natural causes. A monistic naturalist is thus quite skeptical with regard to many folk concepts and many of the everyday beliefs that we seem to presuppose in our human practices. Instead, we should try to find ways of expressing and explicating these beliefs and concepts through the language of science so that we may come to understand these practices more fully. In the words of Sellars, there is then a tension between the "scientific image" and the "manifest image," and monistic naturalists argue that we should favor the former over the latter. Some have additionally suggested that the latter can be reduced to the former. The term "monistic naturalism" is used since it says something particular about this approach; namely that reality should be conceptualized from only *one* perspective, and that is the perspective of modern science. This monistic proposal encompasses ontological and epistemological claims, as well as claims regarding semantics and the nature of higher-level language. It is possible to recognize several religious naturalists whose views on the natural world seem to include monistic elements: Willem Drees, Charley Hardwick, and Loyal Rue.

As described in chapter 1, Willem Drees argues that ontological dualism has been found wanting, and he thinks this is the case given several neuroscientific discoveries made in recent years. He writes, "A dualism of matter and non-physical mental substances is unlikely, since the relation between the mental and the physical is very intimate."[18] Contrary to this dualistic conception of reality, we should concede that "our natural world is a unity in the sense that all entities are made up of the same constituents."[19] Drees calls this thesis "constitutive reductionism."[20] Drees adds to this naturalistic proposal that "the natural world is the whole of reality that we know of and interact with."[21]

Hardwick, by virtue of endorsing physicalism, expresses a similar view to that of Drees when he claims that "only the basic objects of mathematical physics exists and that everything at a higher or more complex level can occur only if there is corresponding occurrence at the level of physics."[22] In this deterministic view, there is no difference in reality without a difference at the physical level.

Loyal Rue, who describes his view as a version of scientific materialism, maintains that "all we have in the real world is matter and its properties," and that all natural facts can be construed in terms of "the organization of matter."[23] Everything that happens in nature and in reality is, according to Rue, contingent on something material in reality. Karl Peters, who I would generally place among the pluralistic naturalists, does have some monistic tendencies. For example, he suggests that naturalism means that "the causes of things are not personal, mental or intentional—except when personal creatures such as humans and probably some animals are involved."[24]

The monistic naturalist seems therefore to hold four ontological propositions to be true:

1. The natural world is all there is, and all entities are made up of the same constituents (*MON1*).

2. If *X* exists, *X* is either something material or a property of matter (*MON2*).

3. The ultimate cause of things is natural, nonpersonal, nonmental, and nonintentional (*MON3*).

And according to Hardwick, who adds a deterministic element to naturalism:

4. Higher-level-happening *Y* corresponds to lower-level-happening *X* (*MON4*).

According to this view, the natural world is all there is, everything that exists is either material or a property of something material and, moreover, deterministic relationships between effects and causes seem to be true. Propositions *MON1*, *MON2*, *MON3*, and *MON4* seem to constitute the ontology of monistic naturalism.

An epistemological monist would, like Drees, hold to epistemological physicalism. This means that "physics offers us the best available description of these constituents [which the world is made of], and of our natural world at its finest level of analysis."[25] Hardwick strongly emphasizes the primacy of physics in claiming that "all truth is determined at the level of physics."[26] And Drees argues that science is the best or perhaps even the only way of acquiring knowledge. Drees and Hardwick therefore make the following *epistemological* claims:

5. Physics offers the best description of reality (*MEN1*).

6. All truth is determined at the level of physics (*MEN2*).

Rue, when writing about the epistemological attitude of a naturalist, states that "what really matters in human inquiry is ferreting out well-justified proposals for belief. And this means evidence."[27] For Rue, then, the obvious epistemology for a naturalist should be some form of evidentialism:

7. *S* should believe proposition *X* only if *X* is well supported or if *X* has been experimentally verified (*MEN3*).

Lastly, with respect to semantics, a monistic naturalist would claim either (a) that we can replace normative language with the language of science, or (b) that nonscientific language is actually meaningless. No religious naturalist seems to adopt either of these positions. On the contrary, religious naturalists tend to emphasize the irreducibility of normative, subjective, and everyday language and commonsense beliefs. But, for the sake of clarity, let us investigate the monistic stance on semantics anyway. Paul and Patricia Churchland are probably the best-known advocates of option (a), and they maintain that prevalent concepts in folk psychology can be replaced with neuroscientific accounts, and moreover that folk psychology, which they take to be a theory of mind, can be abolished.[28] An example of someone who holds to position (b) would be the logical positivist A. J. Ayer. He suggested that nonscientific statements are cognitively meaningless, as they cannot be verified through the methods of science.[29] These claims can be formulated in the following propositional statements:

8. Normative statements X, Y, and Z are reducible to the vocabulary of science (*MSN1*).

9. Normative statements X, Y, and Z are cognitively meaningless (*MSN2*).

These propositional statements regarding ontology, epistemology, and semantics constitute the core commitments of monistic naturalism. We will now move on to consider its counterpart, pluralistic naturalism.

Pluralistic Naturalism

The term "pluralistic naturalism" conveys the view that there is more than one ontological category, more than one way of knowing the world, and different, legitimate ways of semantically construing and conceptualizing reality.

As mentioned earlier (in chapter 1), nonreductive naturalists, whom I refer to as pluralistic naturalists, typically stress the idea of emergence, ontologically suggesting that new properties emerge through the relationship between purely material entities. When one speaks of emergent properties or emergent phenomena, one usually refers to consciousness, culture, morality, free will, and so on.

Stuart Kauffman and Philip Clayton have argued that reality should be construed in terms of higher and lower levels, and moreover that the reductionist view of the world, as construed by monistic naturalists, has become increasingly implausible in recent years. Kauffman and Clayton write that the concept of emergence "presupposes the existence of levels of organization in the natural world."[30] The whole idea that emergent properties are irreducible to lower levels of reality assumes further "that reality is divided into a number of distinct levels of order."[31] For these two prominent thinkers, the following proposition constitutes an important part in their proposal:

1. Reality consists of a hierarchy of higher and lower levels (*PON1*).

This idea contradicts the essential message of *MON1*. Moreover, according to Clayton, emergence theory suggests that higher-level phenomena have emerged from lower levels, and that "more complex units (i.e., emergent properties) are formed out of more simple parts."[32] Similarly, Ursula Goodenough has argued that properties emerge as a consequence of shape interactions. For example, "The interaction of water molecules (nothing but) generates a new property, surface tension (something more)."[33] In this way, higher-order properties emerge by virtue of the interactions between the lower-order constituents.

Thus the following proposition is a part of a pluralistic naturalist's view of emergence:

2. Higher-level *Y* has emerged from lower-level *X* (*PON2*).

Kauffman, furthermore, stresses what is probably the most important aspect of emergence theory, namely that emergent properties/phenomena are irreducible with respect to lower levels of reality. For Kauffman, this emergent reality is clearly manifested in the irreducibility of the biosphere with respect to physics. One of his more famous examples of irreducible organisms is the human heart. The function of the heart is to pump blood. This means, according to Kauffman, that "the function of a part of an organism is typically a subset of its causal features."[34] Kauffman argues from this that even if the physicist were able to deduce all the properties of the heart, he still would have no way to "pick out from the entire set of the heart's properties, the pumping of the blood as the causal feature that constitutes its function," the function that actually accounts for its existence.[35] Thus the emergence of the heart, which requires an evolutionary explanation, is irreducible with respect to physics; in some sense, biology transcends the world of physics.

According to Loyal Rue, meaning itself is an emergent property of nature. Rue argues that we can understand what a living thing is and what its purpose may be by seeing how it carries on. The "epic of evolution," the story about the creation of matter from energy, clearly shows that "what really counts in the game of life" is to "endure and reproduce" so that we may achieve our "biological teloi."[36] For all species, what ultimately matters is living.[37] Consequently, by achieving reproductive fitness one is in this picture living a good and meaningful life. Living is the ultimate fulfillment of life, and the continuation of life is therefore the ultimate and objective value for all life forms.[38] To carry on in our pursuit of reproductive fitness is the ultimate purpose, or grand telos, of life.[39]

Rue stresses the importance of two intermediate goals for achieving the end goal of reproductive fitness: personal wholeness and social coherence. These refer to the creation of conditions such that the construction of coherent and cooperative groups is made possible. By creating and sustaining these conditions, we will maximize the odds of achieving reproductive fitness, which consequently enables us to live a good life. Meaning, from Rue's perspective, is thus objective and "inherent in the objective world by virtue of the various telê found embodied in the heritable traits of living organisms."[40] Given that biological behaviors have emerged from the pointless and purposeless matter of the cosmos, we can appropriately say that teleology is an emergent property of nature.[41]

Emergence theory, on Kauffman's and Rue's view, suggests that:

3. Higher-level Y cannot be reduced to or be replaced by lower-level X (*PON3*).

Propositions *PON1*, *PON2*, and *PON3* seem to constitute the basic elements of modern emergence theory, and the ontology of pluralistic naturalism. They are, one could say, minimal and necessary parts of any emergence theory. However, theories of emergence are typically divided into two forms: one weak and one strong. Proponents of weak emergence not only maintain that reality is constituted by different levels (*PON1*), that higher levels have emerged from lower levels (*PON2*), and that higher levels are irreducible (*PON3*); they also add an epistemological claim to the theory: that we are *epistemologically unable* to deduce higher levels from lower levels. Kauffman (as was seen in chapter 1) argues that the biosphere as a whole fits that description. He describes the epistemological problem of predicting the course and direction of the biosphere and the processes of evolution. There are too many variables to take into account when trying to simulate the outcome of the evolutionary process and what kind of organisms it likely will produce. Science encounters a clear boundary. Kauffman writes that one would have to "carry out infinitely many . . . simulations in order to model our specific biosphere with perfect precision." But, Kauffman continues, "Obviously, no one could get this much time on a supercomputer."[42] The biosphere is creative in ways that are essentially unpredictable.

Gordon Kaufman expresses a similar view when he writes about the *creativity* in the world of nature: "We do not (and may never be able to?) understand the mystery of how greater and more complex things can come out of simpler and lesser things."[43] Indeed, given this clear epistemological boundary, our current explanations for the biological creativity of nature have to be deemed unsuccessful: "None of these descriptions and quasi-explanations, it should be noted, give us any real knowledge of the future toward which this whole creative process may be moving."[44] The emergence of new realities is, in Kaufman's words, a *mystery*.[45]

This seeming unpredictability regarding the processes of evolution is also emphasized by Karl Peters: "Creativity points to a system, the parts which work together in unpredictable ways to create such things as new life, new truth, and new community."[46] Goodenough suggests something similar and claims, "Emergent phenomena are not prefigured. They come for free, apparently out of thin air."[47]

Thus the views of Kauffman, Kaufman, Peters, and Goodenough imply that:

4. We are epistemologically unable to deduce higher-level entities/properties from low-level physical laws (*PON4*).

It seems that propositions *PON1*, *PON2*, *PON3*, and *PON4* constitute what many have called the weak thesis of emergence, a thesis that is mostly committed to an epistemological view of emergent properties.

Those who propose strong emergence, however, add that an emergent phenomenon X not only has to be epistemologically irreducible, it also needs to exhibit ontological newness. This is in philosophical literature also referred to as *novelty*, whereby something ontologically new emerges that cannot be identified in terms of the physical constituents. "New levels of organization bring into existence new kinds of being," as Donald Crosby puts it.[48]

Humans exhibit such novel capacities, for example "symbolic representation," the ability "of the human brain to use symbols (words) to refer to indexes and to sets of indexes, and to use syntax to indicate the relationships of these words to one another."[49] Subjectivity has gradually entered the natural domain through the process of biological evolution. Crosby writes, "The capacity for self-directed selection among alternatives is a function only of complex levels of biological, and especially neuronal, organization."[50] Thus we have radically new properties entering reality.

Novelty is quite often defined in terms of causal effectiveness. Typically, thinkers who are inclined toward strong emergence maintain that an emergent property must play some sort of causal role in order to be considered a genuine feature of the universe. When emergence theorists speak of causal effectiveness it usually involves the notion of *downward causation*, which is the most distinctive feature of strong emergence, but also its greatest challenge.[51] By "downward causation" they mean to suggest that a higher-level entity or emergent phenomenon X manifests causal powers so that X affects its constituents, or the whole causally affects its parts.[52]

Kauffman seems to hold to this view with respect to the biosphere. He claims that specific happenings in the biosphere can alter the "molecular makeup" of the biosphere as a whole, and vice versa. For instance, if a specific biological organism were to go extinct, then that scenario would affect the course of evolution, since the specific proteins, genes, molecules, and so on that were particular to that organism would no longer be present in the biosphere.[53] The extinction of one species could have an immense effect on its biological surroundings. Thus it seems that it is possible to interpret Kauffman as to be arguing for the idea of downward causation, where higher-level entity X (a biological organism) can causally affect its lower-level constituents (the molecular makeup of the biosphere as a whole). This gives us an additional proposition to strong emergence:

5. Higher-level phenomena Y can exert causal efficacy on its constituent parts (*PON5*).

The concept or theory of emergence seems to imply a pluralistic ontology, a world in which there exists a plurality of ontological categories. Clayton calls this view "emergent pluralism," a kind of ontological pluralism that he believes is compatible with a thoroughly naturalistic worldview.[54]

As previously mentioned, a question connected to religious naturalism is that of the authority of science, and whether science is the only path to knowledge. For Kauffman, reductionist thinking leads to severe epistemic injuries, one of these injuries being a publicly accepted dichotomy between the natural sciences and the humanities. As a result of this dichotomy, the humanities have been relegated to dealing with "second-class knowledge." Therefore, research within the humanities cannot be viewed as knowledge-producing at all.[55] Kauffman argues instead that the natural sciences are limited, one important reason being the unpredictability of nature. Hence, with respect to the humanities, Kauffman is convinced that they contain truth as well.

Donald Crosby argues not only for an interdisciplinary pluralism with respect to knowledge but also a richer and irreducible pluralism in the natural sciences. He writes, "Finally, are these other sciences not, or should they not ideally be, reducible to physics as the most basic science?"[56] He answers in the negative. We must, he says, concede the different sciences' common goal of understanding nature, while keeping in mind their different ways and methodologies of doing so.[57] He therefore strongly rejects *scientism*, in his words a secular and "uncompromising faith in science *über alles*, that is in the absolute dominance and hegemony of science over all other disciplines."[58]

Loyal Rue shares Crosby's critical attitude toward scientism. We should be wary of portraying science as the antithesis of religion, and those who wish to hold to this view must be aware that they are going beyond science. For example, people who have claimed that the existence of God has been shown to be illusory by modern science are presenting claims that are not deducible from empirical theories. Such claims can be neither confirmed nor disconfirmed by science, and "science qua science presents no obstacle to theistic belief."[59] Those who insist on claiming that science has disproved the existence of God are making philosophical claims, not scientific claims. They are, as Rue concludes, "confusing science with scientism."[60]

Thus, if we follow Crosby and Rue, we would hold the following propositions to be true:

6. There are other disciplines than science capable of producing knowledge about reality (*PEN1*).

7. There are meaningful claims that fall outside the boundary of science (*PEN2*).

Lastly, there is the question of semantics and whether we should believe in the primacy of the vocabulary of physics, or if we should be pluralists in this area as well.

Kauffman has argued that the greatest weakness of reductionism is its inability to account for teleological language. Essential for teleological language are "means-ends" explanations in which "reasons appear as causes of behavior."[61] Now, the problem for the reductionist is that it does not seem possible to replace teleological language with physical language. Consider, for example, the proposition "Lisa has left her house to go shopping." It does not seem possible, argues Kauffman, to state the necessary and sufficient conditions about human actions in order to reduce the statement to lower-level languages. "The physicist has no way to pick out the relevant subset of events that constitute the action."[62] The tools and methodological resources of physics are not able to sufficiently explicate or reduce the teleological statement "Lisa has left her house to go shopping." Eliminative reductionism fails, which seems to suggest that teleological language is nonreducible and emergent.

Rue, like Kauffman, argues that teleological language is an unavoidable part of reality, something that even the natural sciences cannot give up on easily. The Darwinian account of evolution, Rue goes on to argue, is infused with teleology, and it cannot be reduced to mere functional language. He writes, "Teleological language remains: functional traits are there because in the past they served the *goal* of reproductive fitness. Natural selection provides powerful tools for explaining specific modifications in living systems, but it doesn't offer a complete explanation for the origins of life. Evolutionary theory *assumes* the teleological nature of living systems."[63] On Rue's pluralistic view, reductionism fails to reduce teleological language to physical language since all physical descriptions of reality necessarily employ teleology.

Charley Hardwick, who seems to lean toward semantic pluralism, exemplifies the irreducibility of teleological language with a solar marker constructed by the Anasazi Indians in New Mexico (between AD 950 and 1150). He writes, "From a purely objective standpoint, the configuration of the slabs and the angles of the sun explain physically the patterns of light cast by the slabs. Yet we can recognize them as a solar marker only because their physical states and configuration are realized in a *cultural role*."[64] Hardwick concludes that the property of being a solar marker is emergent with respect to physics. That is, even if we knew all *physical properties*, this would still not lead us to conclude that the slab of rocks constitutes a solar marker. The construction of the solar marker is connected to the intentions of subjective beings and cannot therefore be an object of ontological reduction. Thus in this case the "autonomy of the anthropologist's domain is secure."[65] Explanations used by anthropologists are therefore safe from reductionism in the sense that the vocabulary of anthropology cannot be replaced by that of physics or some other more basic science.

For Drees, a similar argument can be made with regard to subjective descriptions and higher-level explanations. He makes the distinction between descriptive and normative explanations.[66] When someone, for example, claims to be perceptually experiencing some specific object, that person is employing a subjective description that we should concede as legitimate if we have no overriding reason to assume the contrary. And we can accept first-person accounts as genuine without necessarily having to say that these accounts are correct, that is, to say that the accounts are true (or false). If we take this distinction seriously, says Drees, we have no reason to suppose that a naturalistic account of reality implies semantic reductionism, or that we have to think that our subjective ways of comprehending reality are false or not as valuable as some emerging scientific ways of understanding the world.

A naturalist who agrees with Rue, Kauffman, Hardwick, and Drees maintains that:

8. Teleological language is irreducible and emergent with respect to physics (*PSN1*).

9. Some explanations may require vocabularies that go beyond physics, or the natural sciences in general (*PSN2*).

Through this investigation of the ontological, epistemological, and semantic aspects of pluralistic naturalism, I have now provided an overview of the philosophical commitments of this less restrictive version of naturalism, which seeks to rectify some of the immediate problems invited by reductionism.

Summary of Monistic Naturalism and Pluralistic Naturalism

The outlined differences between monistic and pluralistic naturalism may give the appearance that there is a strict line between the two; that you either fully embrace pluralism or monism in this discussion. However, most religious naturalists seem to adopt both pluralistic and monistic elements in their naturalistic projects (with the exception of Stuart Kauffman, who seems to adopt a full-blown pluralistic approach). Charley Hardwick, for example, holds to both a monistic ontology and epistemology yet also affirms the view according to which there are many different, legitimate, and complementary ways of describing reality (he affirms semantic pluralism).

Another example of combining both monistic and pluralistic elements can be found in Drees's proposal. Drees's form of naturalism involves both ontological and epistemological monism. However, he clearly affirms semantic pluralism. But is it coherent to adopt semantic pluralism while holding to both ontological and epistemological monism (or reductionism)? Do not ontology and epistemology carry some implications regarding semantics?

When one turns to Loyal Rue the discussion becomes even more interesting. In comparison to Hardwick and Drees, Rue combines his monism and pluralism not only in the sense that he embraces both these positions with respect to different issues; Rue's ambition is stronger in the sense that he mixes pluralism and monism in his ontology. Rue has stated, as mentioned earlier, that everything existing is either something material, or at least a property of something material. As a materialist he is thus a monist, and he clearly states his naturalism to be a kind of monism.[67] Nevertheless, at the same time he finds strong emergence to be the most effective framework for understanding specific phenomena within reality and reality as a whole, which takes his project in a pluralist direction (what Clayton calls "emergent pluralism").

Conclusions

We have now explicated the core naturalistic commitments of several religious naturalists. My main suggestion is that we should construe naturalism in two forms: monistic naturalism and pluralistic naturalism. The term "monistic naturalism" connotes a view according to which reality should be conceptualized from only one perspective, that is, the natural sciences and especially physics. This view involves ontological and epistemological as well as semantic statements, and is closely associated with a reductionist paradigm. "Pluralistic naturalism," by contrast, suggests that the world is hierarchically structured into higher and lower levels, and furthermore that higher-level phenomena causally affect their lower-level constituents. Moreover, knowledge is not contained only in physics or the natural sciences more generally; there are many disciplines capable of producing knowledge. Regarding semantics, a pluralistic naturalist believes teleological language to be irreducible, that it is not possible to explicate teleological statements in terms of brute physical language.

Most religious naturalists combine monism and pluralism. However, some express both pluralistic and monistic understandings when it comes to ontology. Is it coherent, for instance, to hold both strong emergence and monism to be true, when it seems that they may be at odds with each other, and have often been construed as opposing alternatives? The next chapter will investigate these evaluative questions and the metaphysical grounding problems invited by both monistic and pluralistic naturalism.

3

The Metaphysical Grounding Problems of Monism and Pluralism

We will now move on to examine some of the central issues connected to both monistic and pluralistic naturalism. We have seen that in the area of knowledge, religious naturalists who favor monism emphasize the primacy of science and physics in particular. Physics defines the ontology of reality. The pluralists opposingly maintain that there is a plurality of ways of knowing reality—through science, the methods particular to the humanities, and common sense. This view, as already described, is intimately connected with the popular framework of emergence. In this chapter, I will critically examine these naturalistic positions and ways of making sense of reality.

I will investigate monistic naturalism in relation to the following issues and questions:

1. Monistic naturalism and antireductionism: Have Drees and Hardwick successfully brought out the antireductionist aspects of their physical monism?

2. Monistic naturalism and semantic pluralism: Can a monistic naturalist coherently hold to semantic pluralism and the autonomy of nonscientific discourses?

With regard to pluralistic naturalism, the following questions will be examined:

3. Is downward causation a good match for naturalism, or will such a combination produce a metaphysical grounding problem?

4. What are the implications of the epistemology of emergence theory?

5. Is there a necessary relationship between the existence of emergent properties and a religious/spiritual reality?

Monistic Naturalism and the Issue of Antireductionism

Monistic naturalists, while sticking to a stricter formulation of the naturalistic ontology, seek to avoid the pitfalls of overly reductionist frameworks that might invite eliminativism. Both Willem Drees and Charley Hardwick stress the need of nonscientific vocabularies and normative concepts for making sense of reality. Here I want to argue that Drees's and Hardwick's strategies for bypassing reductionism—that is, their attempt at showing how their respective framework can escape eliminativism and naïve scientism—are philosophically unsuccessful. I hope to bring to light the inherent tension within these broader attempts to steer materialism and physicalism away from sheer reductionism.

As was shown earlier, Drees's belief is that modern science brings with it a view of reality according to which there is no dualism, no spiritual realm, within the natural world. Drees, as we have also seen, confesses that his materialism goes beyond the deliverances of science, but that it nevertheless stays very close to the concepts of science; it is "low-level metaphysics."[1] While Drees opts for ontological reduction within the world, he remains modest concerning the limits of science when it comes to providing an ultimate explanation for the mysterious character of our universe. Limit-questions form an integral part of Drees's religious vision. That is, the sort of "questions that emerge at the limits of scientific understanding."[2] Indeed, for Drees the richness and complexity of reality call for epistemic modesty and an honest agnosticism.[3] It is because of such agnosticism that Drees maintains that his take on materialism is compatible with theism and divine action, albeit in a noninterventionist version articulated through primary and secondary causation.[4]

Drees is honest about the fluid character of his position. It combines elements from religious agnosticism, noninterventionist theism, and a general agnostic attitude concerning the epistemic reach and competence of the sciences. However, due to this creative combination of different ontological and epistemic commitments, it is rather difficult to pinpoint Drees's materialism. Does not this opening to theism, and his affirmation of the intrinsic limitations of the scientific enterprise, undermine his ambition to formulate a materialist ontology? Drees attempts to distance his own, more modest, form of materialism from stricter (and overly metaphysical) ones, but he ends up introducing ontological and epistemic commitments that square poorly with a materialist ontology and affirmation about the nature of the world. I will return to this issue in chapter 5.

Not only is it difficult to make sense of Drees's materialism but it suffers from severe metaphysical grounding problems as well. David Ray Griffin has explored the tension within materialism in a sharp response to Drees, focusing mostly on the issue of human subjectivity and mentality. As Griffin notes, Drees seems to assume the adequacy of materialism but provides little in the way of demonstrating the superiority of materialism over nonmaterialist frameworks when it comes to accounting for consciousness, subjectivity, agency, morality, and similar higher-level phenomena.[5]

With regard specifically to the issue of subjectivity, Drees writes that it constitutes "a major challenge" for the naturalistic description of the human person. This challenge is due, as Drees confesses, to the intrinsic difference between "the experience from within and a description from the outside."[6] Despite the seeming difficulty of naturalizing the phenomenon of subjectivity, Drees remains hopeful that future scientific investigations—coupled with good philosophical analysis—will provide explanatory relief and shed light on how such a complex feature of human personhood has emerged through purely material configurations.

As Griffin further notes, Drees employs first-person language and a variety of agential notions and concepts.[7] Drees suggests that we have a natural capacity for making moral choices and to engage in moral deliberation. This is the ability to reflect on one's past actions "and potential consequences of various options."[8] This is what freedom means, according to Drees. Freedom is self-determination, controlled by rational reflection. Rational reflection therefore plays an important part in moral transformation. The language of "moral deliberation" and "self-determination" suggests that S has the agential capacity to freely act. S is not determined by physical factors, but S is, to some extent, a free creature. S is ontologically capable of making ethical, moral, and spiritual decisions, and to rationally assess potential consequences of these decisions. Drees suggests that even though it has been demonstrated that there are tight links between certain neurological events and our emotional and mental life, determinism does not seem to follow.[9] Freedom, as Drees says, does not conflict "with an evolutionary understanding of ourselves."[10] Agency and moral deliberation are critically important to Drees's form of religious naturalism. *If* he can show that agency and subjectivity are real phenomena and fully compatible with the primacy of physics and the idea that all entities are made up of the same constituents, then Drees would be able to steer clear of a precarious grounding problem for his monistic framework. Thus he would be able to show the coherency of two of his internal commitments, namely agency and ontological naturalism (*MON1–4*).

However, Drees does not justify his materialist views of consciousness and subjectivity. Drees hints at scientific studies that seem to show a significant causal connection between brain events and mental states. Yet, as has often been remarked within the broader field of philosophy of mind, this way of appealing to correlation studies does not explain the nature of subjectivity nor how mental phenomena are hooked up with the physical. As David Chalmers writes with regard to this common fallacy of using neurobiological explanations for tackling the hard problem of consciousness, "None of these accounts explains the correlation: We are not told why brain processes should give rise to experience at all. From the point of view of neuroscience, the correlation is simply a brute fact."[11]

Correlation is not sufficient to ground subjectivity within materialism, nor is it enough to fend off nonmaterialist rivals such as dualism and panpsychism. In fact, a dualist or panpsychist can easily concede the fact of correlations between neural events and human experiencing, and even that there is a relationship of dependency between psychological properties and the physical base level. The mere claim of causal connections between the mind and the body "leaves unaddressed the question what *grounds* or *accounts for* it—that is, the question why the supervenience relation should hold for the mental and the physical."[12] Drees has not adequately spelled out the route for fully naturalizing human subjectivity and freedom. Indeed, I concur with Griffin that Drees's overall reductionist leanings seem to push the phenomenon of subjectivity toward eliminativism. This should not come as a surprise given that it is very difficult to understand mental states as causally efficacious on Drees's constitutive and ontological reductionism. Thus Drees has not provided adequate reasons for thinking that his understanding of materialism takes us beyond eliminativism, and that it can preserve various important higher-level features connected to what it means to be a human being. Monism seems to land us inevitably in a metaphysical grounding problem regarding human subjectivity.

We can spot a similar ontological tension in Hardwick's physicalism, but for slightly different reasons. As we have seen, Hardwick, drawing on the work of philosopher John Post, maintains that a minimal physicalism entails three core claims: everything is ultimately physical; there is no difference without a physical difference; and all truth is determined by truths at the physical level.[13] Physicalism, in its standard version, is usually taken to imply the elimination of nonscientific domains and discourses. Post, however, is honest about the fact that talk "about persons, intentions, consciousness, and the functional states of organisms, for example, seems especially resistant to reduction even of the weakest kind."[14] The obstacles to the reductionist project, as Post notes, are both "numerous" and "severe."[15] Post therefore seeks to uphold a physicalist ontology while avoiding the problems invited by crude reductionism. Post introduces here the idea of *determination*. That is, while the physical level determines everything that happens on higher levels, not all higher-level phenomena can be reduced to purely physical stuff.[16] Thus "determination is an alternative to reduction."[17] What this means is that the physicalist does not have to join the eliminativist in saying that everything can be neatly deduced from purely physical arrangements. The physicalist, in this picture, is *not* claiming that higher-level truths "can be *ascertained* from the physical, but only that they are *determined* by the physical. Given the physical, the rest have to be as they are, but how they are does not have to be ascertainable or predictable."[18] For Post, then, the physicalist is not required to explain the emergence of consciousness, morality, meaning, and so forth. In fact, the "explanation may now and forever belong to some nonphysical domain of discourse."[19] Instead of facing the various metaphysical grounding problems head-on, Post seeks to lower the explanatory burden of physicalism.

Nonreductive physicalism, as defined by Post, is not committed to the stricter epistemological claim that a physicalist needs to explain the appearance of various higher-level phenomena from interactions and combinations at the physical level. It is sufficient to claim that Y is ontologically dependent on X. However, Hardwick's and Post's metaphysical move seems to encounter a similar problem to that of Drees's naturalistic project. That is, it leaves the relationship of determinacy (supervenience) brute and unexplainable. And, as I argued above, Y being ontologically dependent on X is compatible with a host of ontologies, including nonmaterialist versions. Jaegwon Kim explains this point well: "Many diverse mind-body theories accept supervenience; for example, type physicalism, functionalism, epiphenomenalism, emergentism, and the double-aspect theory."[20] Of course, each of these ontologies will provide a different explanation for *why* there is a relationship of dependency. Hence, we should conclude with Kim that "the bare statement that a family of properties supervenes on another does not tell us much."[21]

The problem here is that Hardwick's usage of Post's physicalism amounts merely to a negative description of the nature of higher-level properties and their relationship to their subvenient base, and this strategic move does not address the metaphysical grounding problem of squaring higher-level phenomena with physicalism. We need to know how Y hooks up with X and how X instantiates Y. Yet Post, as shown, maintains that the explanation "may now and forever belong to some nonphysical domain of discourse."[22] Post's clear ambition to uphold the autonomy and irreducibility of "nonphysical discourses" pushes this framework in the direction of ontological agnosticism, which contradicts—and severely so—the spirit of physicalism.

Drees and Hardwick both assert that their specific formulation of physicalism/materialism is different from a nothing-buttery form of reductionism that eliminates consciousness, subjectivity, and meaning. Their respective strategy for taking their projects in a nonreductionist direction should be questioned, as neither Drees nor Hardwick/Post offers satisfactory explanations for how purely material configurations can provide enough ontological resources for the emergence or instantiation of seeming nonphysical properties. As we'll see below, Drees and Hardwick employ another strategy, based on semantic pluralism, for articulating and defending nonreductionism.

Monistic Naturalism and Semantic Pluralism

While both Hardwick and Drees adopt an ontologically reductionist view of the world, they both seek to preserve, in different ways, the irreducibility of nonscientific discourses. Drees affirms, in contrast to the eliminativist Paul Churchland, *conceptual and explanatory nonreductionism*. Churchland's thesis that mentalist language, folk psychology, and higher-level language in general can be replaced by the language of neuroscience (or some other appropriate physical science) seems, according to Drees, "exaggerated" and epistemically unjustifiable.[23] And Hardwick, as we saw, expresses something similar when he states that "physicalism must allow irreducible nonphysical vocabularies to express the properties essential for entities" that belong to domains outside the confines of the natural sciences.[24]

If this notion of irreducibility can be maintained, they argue, then one can avoid a crude scientific imperialism whereby all nonscientific concepts can be reduced to or replaced by the language of fundamental physics. To use my previously outlined typology (in chapter 2), they claim that it is possible to combine *MON1* and *MON2* with *PSN1* and *PSN2*.[25] *MON1* and *MON2* state that whatever exists should be something physical, that is, something that is in principle discoverable by science. *PSN1* and *PSN2*, in contrast, affirm semantic pluralism and the idea that higher-level semantic facts are irreducible to the physical sciences. We must then ask, can an ontological and epistemological framework such as monistic naturalism accommodate semantic facts and normative concepts?

The naturalist, as Huw Price has explained, faces a "placement issue."[26] He asks the following question: "If all reality is ultimately natural reality, how are we to 'place' moral facts, mathematical facts, meaning facts and so on?"[27] Price points out that it remains a mystery how a naturalist could successfully place these facts in a purely natural world. Price is not raising the issue about how we can *derive* semantic facts (for example, moral or mathematical) from nature but rather how physical fact *X* might be both a physical fact and a moral fact. Various normative truths (or semantic facts) seem quite difficult to square with monistic naturalism given that monists believe only natural facts to be real. Thus we encounter a placement problem, or what I have termed a "metaphysical grounding problem."

Terry Horgan and Mark Timmons have described the situation in a similar way. Naturalism is an "accommodation project," meaning that the aim is to accommodate and incorporate discourses (higher-level and normative) into the framework of naturalism.[28] These naturalistic programs, according to Horgan and Timmons, "usually treat the vocabulary of natural science as relatively unproblematic, and various kinds of 'higher level' discourses as needing naturalistic accommodation."[29] They argue, however, that this kind of project encounters problems with regard to semantic discourses. Semantic properties would "be metaphysically extremely queer; they would be quite unlike the naturalistic properties that natural science talks about."[30] If we then adopt a framework that "identifies PROPERTIES with natural properties that do *not* have to-be-pursuedness built into them, then we thereby squeeze out the normativity itself."[31] Hence, given that natural facts are devoid of higher-level meaning, it is hard to metaphysically ground semantic normativity in a world of purely natural facts.

One other option, to avoid epiphenomenalism, is to appeal to a relationship of supervenience. This strategy, as we saw earlier, is unsuccessful, as it leaves the supervenient phenomenon in question unexplained and brute. Claims of supervenience—and claims of ontological dependency in general—do not let us know how the material level instantiates semantic or normative properties. Supervenience by itself turns into a form of ontological mysterianism, and to counter this problem the relationship between semantic phenomena and the physical level of reality needs to be shown to be "*robustly* explainable in a materialistically acceptable way."[32]

Hence, it seems as if *MON1* and *MON2* imply the negation of *PSN1* and *PSN2*; Drees's and Hardwick's attempt at combining or synthesizing ontological monism and semantic pluralism appears rather problematic.[33] One could, of course, respond to this argument by simply allowing the eliminativist to rule out nonphysical vocabularies from the natural order. The eliminativist in question would ask us to bravely embrace the ontological consequences of naturalism and rid the world of those semantic facts that cannot be adequately construed as natural properties. According to the eliminativist, we should simply accept the consequences of not being able to metaphysically ground higher-level phenomena within the natural order as currently described by the natural sciences. Given that Drees and Hardwick resist this solution, monistic naturalism seems to be stuck in a tight spot.

The ontological stance taken by Drees and Hardwick does not allow for semantic normativity and ends up eliminating nonscientific discourses, despite their best intentions to preserve higher-level linguistic practices and avoid a crude reductionism. They have provided some good reasons for thinking that semantic normativity escapes ontological reductionism (see chapter 2), and that it cannot be reduced to purely physical properties. Yet semantic facts seem strangely out of place given the kind of naturalism that both Drees and Hardwick are pushing for in this debate. This is bad news for the coherency of monistic naturalism. This form of monism seems unable to metaphysically ground the normativity required for upholding the reality of higher-level discourses. To conclude, they need an ontological framework that is less restrictive, less reductionistic with regard to what exists, in order to employ normative concepts and preserve the autonomy of nonscientific disciplines.

Pluralistic naturalism is a direct response to some of the problems created by reductive monism and similar ontologies. Pluralistic naturalism is indeed less restrictive and offers some interesting ways of bringing together the religious and naturalistic components of religious naturalism. However, as we shall see, pluralistic naturalism faces other sorts of metaphysical grounding problems.

Pluralistic Naturalism and Emergence Theory

Pluralism is by far the most attractive approach to ontology among proponents of religious naturalism. A pluralistic naturalist holds to a number of beliefs pertaining to the "layeredness view of reality." This thesis entails that, ontologically speaking, reality consists of several irreducible levels. Furthermore, it claims that science is not the only path to knowledge, but that other disciplines, corresponding to different ontological levels, can give us insight into the nature of reality. A pluralist also maintains that higher-level languages are semantically irreducible (e.g., teleological language). This form of ontology is important for religious naturalists, as it enables them to appreciate nature religiously and spiritually, while staying true to the philosophy of naturalism, that is, without introducing dualism or supernaturalism.

I have already suggested that pluralistic naturalists take reality to contain different emergent properties/phenomena and levels. Emergence theory is committed to nonreductionism, meaning that higher levels are nondeducible and that they are causally effective. In order to fully understand what it means for something to be emergent, we must therefore probe deeper into some of the central concepts of this theory, concepts such as *nonreducibility, unpredictability, novelty,* and *downward causation.*

Emergent Phenomena as Unpredictable and Unexplainable

One of the central ingredients of emergence theory is that emergent phenomena are unexplainable and unpredictable. What do emergence theorists have in mind when they claim that emergent properties are unexplainable?

David Chalmers suggests that an emergent property, a higher-level phenomenon, can be strongly or weakly emergent with respect to its physical base. A higher-level phenomenon is strongly emergent when we cannot, not even in principle, deduce the truths of that phenomenon from the truths of its lower-level base. Weak emergence, in contrast, means that emergent phenomena are *unexpected* "given the principles governing the low-level domain."[34] Weakly emergent phenomena are unexpected but not unexplainable. Chalmers argues that you could very well maintain weak emergence without invoking strong emergence, but it is often the case that strong emergence implies weak emergence: "It often happens that a high-level phenomenon is unexpected given principles of a low-level domain, but is nevertheless deducible in principle from truths concerning that domain."[35]

By saying that an emergent property/entity/level is unexplainable, one may think that such phenomena are ontological and epistemological mysteries. Stuart Kauffman seems to affirm such mysterianism. On his view, the direction of the biosphere is beyond computational capacities. Even with access to a supercomputer, we cannot simulate the outcome of the evolutionary process; it is literally a mystery. Because of this some suggest that even if we knew all the relevant facts of the lower levels of reality, we would still not be able to deduce an emergent property/entity. In this way, emergents are strongly epistemologically independent of their ontological base.

Philip Clayton, on the contrary, thinks that even if we cannot tell the whole story about emergent phenomena by studying the base structure from which they emerged, we might at least be able to say something. Clayton writes, "One can sometimes predict the emergence of some new or emergent qualities on the basis of what one knows about L_1. But using L_1 alone, one will not be able to predict (i) the precise nature of these qualities, (ii) the rules that govern their interaction (or the phenomenological patterns), or (iii) the sorts of emergent levels to which they may give rise in due course."[36] Contrary to Kauffman, Clayton believes that the sorts of emergent properties that may come into existence are not *completely* beyond human understanding.

If we are saying that emergent properties are unpredictable, are we therefore also saying that they are unexplainable? It seems that many emergence theorists would answer in the affirmative. However, affirming unpredictability, that they are nonreducible, does not mean that they are beyond causal description. Claus Emmeche, Simo Køppe, and Frederik Stjernfelt state that we must "differentiate between description and explanation. To describe an event is not an explanation. In continuation of this, we can distinguish between a descriptive causality and an explainable causality."[37] Thus we may still be able to describe the coming into being of new emergent phenomena, although we have no explanation for *why* they came into being.

The idea of these phenomena as unexplainable is intimately connected to the concept of novelty, or the "newness" of emergents. Jaegwon Kim reasons as follows: "I believe that 'new' as used by the emergentists has two dimensions: an emergent property is new because it is unpredictable, and this is its epistemological sense; and, second, it has a metaphysical sense, namely that an emergent property brings with it new causal powers, powers that did not exist before its emergence."[38] We will now proceed with a closer look at this crucial concept of novelty.

Emergent Novelty

Emergentists stress the newness of emergent properties, which usually means that increasing complexity on lower levels gives rise to properties or entities not previously found in nature. This is not just an epistemological assertion but also signifies something ontologically new. One very common way of expressing the idea of novelty is to say that the whole is more than the sum of its parts. Novelty, from an evolutionary perspective, means that, "in the course of evolution, exemplifications of genuine novelties occur again and again. Existing building blocks develop new configurations; new structures are formed that constitute new entities with new properties and behaviors."[39] It is worth noting that some emergentists have a stronger sense of novelty than others. For some, emergent phenomena are an entirely new kind of thing, not found even in potentiality. For others, what is new existed in a latent form, which suggests that "all things and all facts are 'pregnant' with whatever may arise from them."[40] New configurations of lower levels, which constitute new relations at the physical level, bring about new entities that possess novel *behaviors*. The word "behavior" often refers to new causal powers, which is frequently framed within the metaphysical theory of downward causation.

Downward Causation

Causal powers, on the emergentist view, are often connected to the philosophical position of "downward causation," which is perhaps the most controversial idea proposed by emergence theorists. It is controversial, since it could possibly cause some tension between the monistic commitment of naturalism and emergence theory, or pull naturalism in the direction of dualism. "Downward causation" means "the process whereby some whole has an active non-additive causal influence on its parts," or the ability of an emergent structure to causally influence its constituents.[41] This seems to be true, as Nancey Murphy points out, when one looks at systems that are influenced by their environments. She describes this as follows: "The system S interacting with its environment is a higher-order system S'. If S is affected by its environment, then this is downward causation from S' to S."[42]

One reason why several emergentists are attracted by the notion of downward causation is that it is easier to justify the ontological realness, and hence the novelty, of emergent properties/phenomena/levels if they actually make a causal difference in reality; that is, if they did not exist, reality would be different from how we experience it. The novelty of emergent properties is therefore highly dependent on the causal capacity of the properties in question. The importance of downward causation lies in showing how emergent properties can be considered real.

Mere irreducibility is not sufficient for upholding the doctrine of emergence and the idea of mental properties as novel. As Jaegwon Kim has argued, being irreducible does not establish an emergentist ontology; it says only "it's not reducible," and so it provides only a negative account of emergence.[43] An emergentist might say that we cannot reduce mental properties to electrochemical events, that it is not possible to reduce normative language to physical language. Even if that is true, it does not establish an emergentist ontology, as it shows only that a reductive explanation of certain phenomena fails.

Thus emergence needs something more, and herein lies the importance of downward causation. An emergent property, in this view, means that new causal powers have come about, causal powers not found in the microphysical components of an emergent property. It is not only impossible to reduce the mental to the physical, but the mental carries certain causal powers not identifiable in the microphysical makeup of the mental. This is what it takes to establish the novelty of emergence. By affirming downward causation, an emergentist can say something positive about emergent properties. This point will be returned to later on.

If we can identify specific instances of downward causation, we might be justified in affirming the existence of emergents. If we can see that Y, some kind of entity, exerts causal influence on physical base X, then we should presumably conclude that Y actually exists. Thus a pluralistic naturalist who subscribes to the ontology of emergence should also affirm downward causation, given that this notion protects emergents from the threat of reductionism. As we shall see, downward causation raises some severe problems for the committed naturalist, and hence for the project of religious naturalism.

Some Grounding Problems for Pluralistic Naturalism

Pluralistic naturalism is at first glance a very promising approach. Contrary to its monistic rival, it seems to have the resources for incorporating and doing justice to higher-level properties, such as consciousness, morality, and values. These are by many thinkers considered essential ingredients for a religious worldview. An additional promise of pluralistic naturalism is that it might help to bridge the gap between the manifest image of reality and a scientific worldview. Hence, it is not surprising that many religious naturalists seek to employ an emergentist ontology in order to demonstrate the possibility of living a religious life as a naturalist.

Notable historical figures have emphasized the religious potentiality of emergence theory. Niels Henrik Gregersen writes, "From the outset, the idea of emergence was embraced by religious thinkers of various kinds. Samuel Alexander developed a position which lingered between a Spinozistic pantheism and an evolutionary theism; Lloyd Morgan combined his naturalistic view of emergence with a classical theism, claiming that all-that-is ultimately depends on God as an 'immaterial source of all.' "[44] Despite the promise of pluralistic naturalism and emergence theory, I would like to raise a few critical questions concerning the combination of downward causation and causal closure; the epistemology, or mysteriousness, of emergence theory; and the relationship between a religious reality and emergent properties.

Downward Causation and Monism: A Good Match?

The pluralistic naturalists under consideration construe emergence and downward causation against the background of a general form of monism;[45] nature is all there is, and there is one reality (monism), although nature is layered (emergence). Pluralistic naturalists are therefore monists in the general sense, without adopting *monistic* naturalism. Monism, in its naturalistic version, is often associated with the idea of *causal closure*, meaning that there are no nonphysical or supernatural causes in nature, and if you trace an event back to its origin, you never have to leave or go outside of the natural order. To reject this idea is to say that there may be an extranatural, nonnatural, or supernatural realm that exerts causal influence on this world. This should be an unacceptable conclusion for monistic religious naturalists and pluralistic religious naturalists alike. In this way, causal closure is a prerequisite for any real form of naturalism. David Papineau, a well-known defender of philosophical naturalism, has argued persuasively for the close connection between naturalism and causal closure, and that nonphysical causes are unacceptable within a naturalist framework.[46]

I will argue that the notion of downward causation breaks the naturalistic commitment to causal closure, and thus introduces dualism. In this way we can see that the pluralist notion of downward causation constitutes a form of metaphysical grounding problem for the naturalist. The origin of emergent properties seems strange in a naturalistic worldview since it is hard to conceive how a purely physical structure can give rise to nonphysical emergent properties.[47] These properties cannot be identified or metaphysically grounded in nature. Moreover, these properties are ontologically independent; hence it is suggested that it is appropriate to ascribe causal powers to these properties. If emergent properties, however, *were* dependent on their physical base structure, then their causal powers would be epiphenomenal. This is because what we perceive to be causal powers on the emergent level are merely produced, and thus determined, by the physical level. Thus downward causation would have to be rejected. This is what the philosopher of mind and emergence critic Jaegwon Kim argues.[48] Kim notes that it is crucial for emergent properties to possess causal powers of their own. Emergent properties that lack causal powers—and whose causal contributions can be sufficiently explained in terms of the relations of the base level—will soon be abolished. However, the notion of genuinely new causal powers of an emergent property P, that is ontologically dependent on base level B, does not sit well with naturalism. The problem is that if we affirm that B can sufficiently explain P, then B must sufficiently explain the causal powers of P. This, however, seems to compromise the causal contributions of emergent phenomena, since anything they seem to cause can be, and is, reducible to a physical cause.[49] The problem, then, is that emergent properties become mere epiphenomena.

I suggest that for downward causation to make sense, the causal powers of emergent properties, such as mental properties, must be located in something nonphysical. This is one way to save emergence theory from the threat of epiphenomenalism and to avoid the grounding problem in question. Of course, this would mean that downward causation now contradicts the principle of causal closure. If downward causation is true, and there are instances of downward causation in the world not reducible to something physical, then we have reasons to believe that there exists something nonphysical. Thus pluralistic naturalism collapses into dualism. The dilemma for the pluralistic naturalist is this: either bite the bullet and embrace the epiphenomenal character of emergent entities, or avoid epiphenomenalism by introducing a nonphysical realm from which emergents draw their causal powers.

The latter approach would come close to William Hasker's emergent dualism. Hasker starts off by challenging the metaphysical idea of causal closure. This principle, which states that "all physical events have physical causes," argues Hasker, is "a requirement for all materialists without exception."[50] The problem facing the materialist, however, is that this principle is not able to "do justice to human action and in particular to human rationality."[51] In a Kim-like fashion, Hasker argues that the idea of reasons as causes does not fit the physicalist worldview, as reasons can never be considered as the cause of actions, or given any role to play in bringing about changes in behavior. Thus "on the assumption of the causal closure of the physical, no one ever accepts a belief because it is supported by good reasons."[52] The reason for the acceptance of a belief, following the idea of causal closure, can be explained solely in terms of physical causes. Hasker saves emergence from a potential metaphysical grounding problem by rejecting the principle of causal closure.

Instead of adopting mere property emergence, Hasker proposes the notion of an "*emergent individual,* a new individual entity which comes into existence as a result of a certain functional configuration of the material constituents of the brain and nervous system."[53] Hence, Hasker's emergence is a form of "subject emergence," an immaterial subject of experience, as opposed to property emergence. In this way, our self is located in a nonphysical substance that is causally effective in the world. This form of dualism is still less dualistic, and so less problematic, than Cartesian dualism, as the emergent individual arose from a physical base.

Hasker's emergent dualism might be able to solve the problem of the causal efficacy of emergent properties. However, I do not think that this form of emergence will be appealing to many pluralistic naturalists, as it seems to stray too far away from a naturalistic worldview. We have here a conflict between some of the philosophical commitments of pluralistic naturalism. Namely, the idea regarding the causal efficacy of emergent properties (*PON5*) does not seem compatible with the necessary naturalistic belief that our universe is causally closed and devoid of nonphysical causation.

Downward causation (DC), as a core doctrine within the emergentist-naturalist worldview, is muddled and obscure. One significant reason for this situation is that the term "to cause" is seldom defined, despite being frequently used within emergentist literature. This issue has been explored by Menno Hulswit. Drawing on a discussion by Emmeche, Køppe, and Stjernfelt, Hulswit brings out the ambiguities and conceptual confusion regarding the issue of causation in higher-level systems, and the precise causal relationship between emergent phenomena and lower-level processes. As Hulswit explains, there are two distinct ways of framing the ontological significance of DC. First, one can frame DC in terms of concrete entities or events. In such a model, which was expressed by the neurophysiologist Roger W. Sperry, mental events exist "in their own form different from neurological events, in endowing subjective phenomena with causal influence and in placing mind in a control role above matter in the brain."[54] Sperry pictures the emergentist causal relationship as that of between two distinct entities. Kim, as we saw above in his critique against DC, describes emergent causation in a similar way when he outlines the problem of mental properties acting efficaciously on their physical base structure. As Kim points out, the idea that the whole can causally affect its lower-level constituents on which its very existence depends seems rather strange. This way of construing DC therefore leads to severe ontological problems. Second, one can understand the causal relationship in terms of general principles or boundary conditions, which basically means that the behavior of lower-level phenomena is *constrained* by higher levels.[55] This way of framing the causal relationship might escape some of the immediate problems of understanding DC in terms of concrete entities; however, it introduces other ambiguities. First of all, the ontology of these general principles remains dubious. How should we understand the realness of these principles? Second, how should we understand the causal relationship between general principles and lower-level events? Do these general principles cause, or bring about, lower-level events? Or is it the case that lower-level events happen *according to* certain general principles? If we take the latter approach, then it seems difficult to understand the higher-to-lower relationship in terms of causality.[56] The underlying problem in theories of downward causation, as Hulswit concludes, is "that in the end nobody really knows what is meant by 'causation' or 'cause' or 'causing.' "[57]

Some have argued that the problems of grounding downward causation within monism, or generally of making sense of the core emergentist claims, stems from the prevailing assumption of metaphysical realism. The pragmatist thinker Sami Pihlström notes, "Most emergence theories seem to be based on a strong scientific or metaphysical realism."[58] Thus the main ambition of the metaphysical emergentist is to investigate "whether there *really* are emergent properties (however they are defined) in the basic structure of the world itself, independently of our conceptualizations of the world."[59] Strong emergentists, as we saw earlier, attempt to frame emergence theory through downward causation. As we also saw, this metaphysical strategy is based on a causalist assumption, meaning that if something is causally efficacious then it enjoys a positive ontological status. Pihlström concurs with Kim that downward causation is metaphysically ill-suited for physical monism, and contradicts the idea that the universe is causally closed.[60] Pihlström's recommendation is that we consult philosophers within various nonmetaphysical traditions, such as Ludwig Wittgenstein, Georg Henrik von Wright, Hilary Putnam, and John McDowell, in order to nonreductively secure human mentality within naturalism. This pragmatic move does not view emergent properties as existing independently of human conceptualizations. Instead, it is more beneficial, Pihlström goes on to argue, "if we considered emergence as a feature of our linguistically structured, practice-involving world in which we, as reflective creatures, think and act through our capacity of normative evaluation."[61] Rather than engaging in the metaphysical projects of naturalizing higher-level properties, we need to simply acknowledge the unavoidable fact that we need to operate with two *standpoints* (Pihlström draws here on Kant) to make sense of our own humanhood and our place in our natural environment. That is, we have to simultaneously think of ourselves as "causally determined elements of the natural world" and as "free, responsible moral agents . . . whose rational capacities and deliberations cannot" be fully described through the concepts of science.[62] This dualism between perspectives accepts the naturalness of both perspectives, important as they are for us to understand human life and the normative notions of freedom and agency. In this way, Pihlström suggests that we can retain the normativity of human mentality without having to speculate metaphysically about downward causation and its relationship to the physical. According to this more pragmatic strategy, it is unnecessary and philosophically problematic to engage grounding problems in a *metaphysical* manner.

Can this pragmatic reformulation of emergence alleviate the problems of pluralistic naturalism? Pihlström wants to avoid the metaphysical problems associated with downward causation, but his nonmetaphysical pragmatism invites a strange kind of quietism regarding the nature and origin of human mentality. Pihlström does not deny that human mentality is an emergent product. He writes, for example, that "agency and normativity surely emerge from mere factuality" and that the phenomenon of normativity constitutes a "qualitative leap."[63] He is, nevertheless, reluctant to ontologically address the issue of human mentality, or attending to the issue of how simple physical configurations can give rise to causally efficacious creatures. On this proposed pragmatism, human mentality is a natural phenomenon that requires no naturalization. Pihlström's account seems to be indiscernible from weak emergence (*PON3* and *PON4*), an epistemic stance that is fully consistent with a more reductionist anthropology. Indeed, Pihlström acknowledges the deterministic implications of naturalism, but he insists that we need particular higher-level concepts to make sense of human nature (he seems to affirm *PSN2*). Pihlström's pragmatism is too weak to establish the irreducibility of human mentality, as we want to know how a "qualitative leap" is possible given the available resources of naturalism. The only way to retain the ontological status of emergent properties is to demonstrate how physical configurations can give rise to normativity, mentality, meaning, and similar features of reality and human life.[64] A quietist dismissal of explaining downward causation will not cut it, but one is forced to positively address the metaphysical grounding problem of human mentality.

Problems with Mystery Emergence

The concept and sense of mystery has a central place in religious naturalism. Emergence theory, according to Stuart Kauffman, takes us beyond the "Galilean spell" of reductionism to the "unknown mystery into which we must live our lives."[65] We do not know what the future holds, but we must, says Kauffman, have faith in God, the god that is the unfolding of nature. The epistemology of emergence theory, as stated in *PON3* and *PON4*, means that it is not possible to reduce Y to its lower-level (or an underlying physical structure) X, nor is it epistemologically possible to deduce the reality of higher-level phenomenon Y from lower-level physical laws. Nature, in the emergentist view, is creative. This creative aspect of nature is something that religious naturalists maintain should motivate a religious response of awe and wonder toward nature and its continuous unfolding. It is suggested that the theory of emergence may show how it is possible to reconcile naturalism and religion. Creativity, in Gordon Kaufman's view, is "somehow emerging—mysteriously, without explanation—between the order and the disorder, the information and the noise, found in every system or structure."[66] Our cosmos is "an evolutionary one in which life and new orders of reality come into being in the course of exceedingly complex temporal developments."[67] Creativity happens, and this "is an absolutely amazing mystery."[68] God, in this emergentist view argued for by Kaufman, is in fact the ultimate mystery.[69] Emergence, and in particular the unpredictability thesis of emergence, is thus used in a number of ways for developing a religiously naturalistic alternative to both full-blown secularism and supernaturalism. We can see that religious naturalists claim the following benefits regarding emergence. First, emergence helps to break the reductionist spell that has often been associated with naturalism. Second, emergence helps to establish a sense of mystery with respect to the creativity of nature. And third, emergence helps to establish an account of faith and a model of God consistent with the framework of naturalism.

According to pluralistic naturalism, emergent properties are nondeducible from their material base, and we have no way of knowing how they came to be. This should invite a sense of mystery, and religious naturalists argue that this may show how it is possible for a naturalist to adopt a religious attitude toward nature and its processes. The religious attitude of this naturalism is thus intimately connected to the epistemology of emergence theory. At the same time, emergence theorists maintain that emergents (for example, mental properties) are causally effective, that by downward causation they affect the material base from which they have emerged. Hence, with increasing complexity, mind, or mental properties, has gained novelty and has become ontologically distinct from its base properties. Because of this novelty, emergence theorists also strongly reject epiphenomenalism. This claim, when scrutinized more closely, does not seem to make much sense. If an emergence theorist believes emergent property Y to be unexplainable, it would seem that she cannot give an ontological account of the causal relationship between Y and its material base X, and hence cannot affirm that Y is ontologically independent. Consequently, the threat of epiphenomenalism increases for the emergence theorist.

How can a proponent of emergence maintain that these properties are causally effective and ontologically irreducible when it is not possible to provide a causal account of these properties given the epistemology of emergence? The fact that these properties, according to emergentism, are unexplainable should not lead one to embrace the strong metaphysical notion of downward causation; one should instead become agnostic concerning the ontology and causal novelty of these properties. The epistemology of emergence, far from inviting the idea of downward causation and strong emergence, actually *prohibits* one from making ontological judgments concerning higher-level properties. According to such an epistemology, it is impossible to tackle the metaphysical grounding problems that the naturalist faces. Indeed, the whole notion of *higher-level* properties becomes problematic, since the emergentist is not able to affirm any ontological distinction between what we perceive to be higher and lower levels of reality. The idea that physical reality is layered assumes that we can make ontological judgments, but this is what the epistemological thesis of emergence makes impossible. Moreover, given the epistemology of emergence, the emergentist is unable to fend off the threat of epiphenomenalism. An epiphenomenalist would argue that what seem to be instances of downward causation are really caused by physical events at lower levels; that, for example, mental events in the brain have no effect on the body or any physical event. Given the epistemology of emergence theory, the emergentist has no way of showing the epiphenomenalist argument to be false. This is the problem for the pluralistic naturalist: what provides a sense of mystery to naturalism, and a religious dimension, undermines the prospect of affirming a richer view of reality in terms of irreducible levels.

My suggestion is that what gets emergence theory into trouble is the idea that emergent entities are mysterious phenomena; we have no way of knowing how they came into existence. Thus the emergentist can never provide a positive account of emergents. One could, in response to this, argue that we will, if given enough time, be able to carry out the necessary computations for understanding how material base X produced property Y. That is, we should not be too quick in ruling out a possible explanation for how emergents have come into being. After all, science develops all the time, and we should not too easily set up boundaries for the reach of science. However, if one maintains that emergents really are predictable, then the novelty of these phenomena/properties will be called into question. As Gregory Peterson writes, "In either case, novelty is taken to be the opposite of predictability; what makes an object or law emergent is, in part, the unpredictability of that object or law in advance."[70] If we deny unpredictability, then we may also have to deny novelty, and thus strong emergence collapses. If we can predict it, we can probably explain it, and if we can explain it, we can probably reduce it; the epiphenomenalist threat returns once again.

Moreover, denying the unpredictability view of emergent phenomena may not be an attractive option for a pluralistic religious naturalist. As argued for above, the unpredictability thesis is quite important for the project of religious naturalism. It makes possible the theological project of developing a nonreductionist and naturalistic conception of God. Nature, if we take the emergentist picture of reality seriously, is creative in the sense that it gives rise to new realities, which, according to Rue, justifies hope and a form of cosmic optimism: life is not determined; it may change for the better.[71] And for Kaufman, Kauffman, and Peters, the reality of unpredictability is intertwined with the notion of Creativity, the amazing mystery of nature. If we give up the idea of unpredictability, then novelty will have to be denied as well. We are left with no novelty, no cosmic optimism, and no creativity. The religious aspect of pluralistic naturalism would be lost.

Is There a Necessary Connection between Emergent Properties, Values, and a Religious Reality?

For a religious naturalist, there is a deep connection between emergent properties and a religious conception of reality, although this connection is expressed differently from one naturalist to another. For example, some naturalists base their religious affirmations on the observation that the biosphere seems irreducible, in the sense that it is not possible to deduce high-level properties from low-level physics. According to Kauffman, we are not able to predict the course of the biosphere.[72] Thus the biological world seems emergent with respect to physics. Moreover, these naturalists have argued that reductionism fails, given its inability to account for teleological and intentional language. Emergence allows for agency, meaning, and values to enter the world. Hence, pluralistic naturalists suggest that emergence provides a desired religious interpretation of the workings of nature. However, is there a connection between the existence of emergent properties and some kind of religious reality? How should we understand this connection?

If one interprets these arguments realistically, then, they seem to encounter some problems. Religious naturalism would in this case be an ontological thesis according to which there are independently existing religious aspects of the natural order—there are some objective aspects of nature that it is appropriate to consider religiously significant, independent of anyone's judgment. Consider Stuart Kauffman's and Gordon Kaufman's view of God as creativity, which they have referred to as the coming into being of new realities and possibilities. Since God is identified with the evolutionary process, the course of which we cannot predict for epistemological reasons, God also becomes a mystery.[73] However, as some religious naturalists rightly note, the word "creativity," let alone "God," is not used by scientists when describing emergence theory. It is, one should say, an extrascientific interpretation of empirical data, or a metaphysical add-on. Ontologically speaking, it seems that the terms "God" or "creativity" do not contribute anything to our understanding of nature. It almost seems as if these terms are ontologically superfluous. Moreover, it is rather unlikely that these philosophical or theological interpretations can be derived deductively from emergence theories as presently stated. From a realist perspective, religious naturalism appears to offer an ontology that goes beyond emergence theory, an ontology that cannot be derived from such nonreductive theories.

Would such a realist interpretation render religious naturalism completely empty of force? The answer to this question depends on how we view the arguments put forward by religious naturalists. If construed as a *positive* argument, meaning that there is a necessary connection between the reality of emergence and religious aspects of nature, it would have to be considered unsuccessful (for the reasons stated above). It does not seem to be possible to deduce these philosophical and theological interpretations from emergence theories as presently conceived. One who studies emergence theory and who comes to recognize the positive ontological status of emergent properties is not deductively, or empirically, forced to adopt a religious conceptualization of the same properties. Something would indeed be missing if these types of arguments were taken as arguments in the positive sense. It would perhaps be better to understand the connection in terms of a *negative* argument.

A negative argument would in this way be an argument that shows that it is possible that Y is true by virtue of X being true, without there being a necessary relationship between Y and X. With respect to the discussion of emergence, one could interpret the argument as saying something like this: By showing the falsity of reductionism and by demonstrating the positive status of an alternative ontology, namely emergence, the reality of a religious worldview becomes ontologically conceivable. There is, then, no necessary connection between the ontology of emergence and a religious worldview, but the former makes the latter at least possible. This would seem to be the most charitable approach to the discussion about how to understand the relationship between emergent properties and a religious conception of reality, if we take the metaphysical background theory to be realism.

Conclusions

This chapter has highlighted several metaphysical grounding problems for both monistic and pluralistic naturalism. I critiqued monistic naturalists, such as Charley Hardwick and Willem Drees, for failing to differentiate their own monism from eliminativism. I also critiqued their attempt at joining ontological reductionism with semantic pluralism. They cannot uphold the autonomy of nonscientific disciplines and discourses.

Pluralistic naturalism is certainly more promising. However, it was seen that the notion of downward causation is not compatible with the monist idea of causal closure. Moreover, it was argued that the idea of emergent properties as being unpredictable and unexplainable invites ontological agnosticism, which in turn undercuts the belief in ontological emergence.

I concluded the chapter with a discussion regarding how to understand the relationship between the reality of emergent phenomena and the possibility of interpreting emergence theory religiously. While the connection between emergence and a religious appreciation of nature cannot be construed in *necessary* terms, it was seen that a nonreductionist ontology such as emergence provides a negative argument for the plausibility of a naturalistic religiosity.

4

The Religious Dimension of Religious Naturalism

Religious naturalists conceive of religion in a variety of ways. Due to the fact that these naturalists have different starting points, it is not very surprising that there is a great deal of diversity concerning the religiosity of their proposals. There is no agreement or consensus between religious naturalists on what it means to adopt a "religious" attitude or what it means to live religiously in a purely natural world. This chapter will explore more deeply the religious component of this naturalistic vision by focusing on three different approaches: religious realism, religious antirealism, and religious pragmatism.

Religious Realism in Rue and Crosby

According to Loyal Rue, religion can be explained naturalistically through evolutionary concepts and ideas. Religion, although culturally complex, is not a sui generis phenomenon that somehow defies scientific explanations. Religion, like other comparable cultural entities, has emerged in the natural course of evolution and is therefore susceptible to evolutionary explanations.

The overarching theory proposed by Rue is of religions as *mythic traditions*. Such mythic traditions offer a narrative account of cosmology and morality, and the joining of these two is achieved through a root metaphor, for example, "God," "Nature," "the Dharma," and so on. Moreover, in order for a religious tradition to survive and flourish, it must utilize certain strategies. Rue suggests five specific "ancillary" strategies that determine the successfulness of a tradition: intellectual, experiential, ritual, aesthetic, and institutional.[1] I will explain these ancillary strategies below.

Religious myths require interpretation in order to appeal to people from different age groups and cultural settings. An important *intellectual* task, therefore, is to interpret religious texts so that religious doctrines become accessible and relevant for everyday life. Rue notes regarding this: "In some traditions the intellectual dimension becomes dominant, revealing a presumption that the religious life is essentially either a matter of believing a set of formal doctrines or a matter of upholding a rigid code of moral conduct."[2] The religious life is also *experiential*, characterized by certain types of experiences that can broadly be construed as "religious," although it is difficult to specify the conditions for such experiences. Rue's main point is that religious experiences are conducive for carrying out the task of a mythic tradition and uphold its authenticity. This experiential strategy, Rue argues, is also important for fighting off religious nonrealism: "If you have once had a mystical, numinous, or visionary experience of divine reality, then the probability of your becoming a religious skeptic or non-realist is reduced to zero."[3] I will come back to Rue's critique of nonrealism shortly.

Closely related to the experiential dimension of mythic traditions are *ritual strategies*. Rue also uses the term "practice" to refer to this ritual aspect of religion. In the broadest sense, Rue writes, "a ritual might be construed as any repeatable unit of behavior, the performance of which engages individuals or groups in the meanings of a religious myth or is conducive to a religious experience."[4] In a similar way to the experiential strategies, the function of rituals is to "reinforce realism about the myth."[5] Indeed, according to Rue, the "very public nature of ritual performances often has the result of enhancing a realist stance towards the myth."[6]

For a ritual practice to function properly, for rituals to invoke a certain form of religious attitude, the aesthetic aspect is critically important. The *aesthetic strategy* refers to the function of art and creativity in the religious sphere. In the area of religion, as well as outside the religious domain, art enables people to "express emotional feelings in uniquely satisfying ways."[7] Furthermore, art helps to convey religious meanings that go beyond religious doctrines. It can arouse a sense of shame, compassion for other people, and gratitude toward the ultimate reality.[8] In the end, art "revitalizes the power of myth."[9]

Finally, a mythic tradition is characterized by an *institutional strategy*, by which Rue means the intention and operation of transmitting a particular myth to future generations.[10] Institutional authority is crucial for the well-being of mythic traditions, and it extends into the spheres of interpretation, ritual, and aesthetic expression.[11]

Rue concludes that these five strategies, none of which can be considered as the "essence" of the mythic, can be seen as overlapping dimensions that together shape and modify the religious life. Religion is a complex phenomenon, and there is no way to locate the core of religion. Yet Rue maintains that it is possible to "reduce the complexity to a pattern of elemental components" and so reveal the "anatomical structure of religion."[12]

Taking this naturalistic project further, Rue suggests (as will also be critically discussed in the next chapter) that the function of religion is to support personal wholeness and social coherence. These twin goals are necessary ingredients for attaining reproductive fitness, which, according to Rue, is the ultimate goal of human life. In this way, Rue concludes that religion is not about God but about *us*.

However, Rue argues that even if traditional religion can no longer be construed in terms of referring to a supernatural reality, the realist intention can still be maintained. Certainly, for a mythic tradition to be effective it needs to be based on a realist understanding. Rue writes, "If my realism about the root metaphor of a myth is compromised, then the fusion of reality and value is compromised; if my sense of the objective reality of certain prosocial values is compromised, then the linkage of those values to my self-esteem will be compromised; if my self-esteem is de-linked from certain values, then the power of these values to command a hearing in working memory will be compromised; and if the values in question fail to gain a hearing in working memory, then they cannot influence the manner of my appraising and coping with any business at hand."[13] Hence, Rue's proposal hinges on a realist understanding of religion, and he therefore seeks to defend religion from what he calls "creeping non-realism." Rue identifies two sources for the increase of nonrealist interpretations: modern science and religious diversity.

Modern science, according to Rue, has interfered with the effectiveness of mythic traditions. This does not mean that science per se has disproved the core beliefs of these traditions, including the long-held belief in a transcendent creator. To the contrary, Rue argues that these questions lie outside the boundary of the empirical sciences. Nevertheless, science has come to weaken the explanatory power of many religions. The mysteries of nature, from a religious point of view, were often subjected to what Rue refers to as "personal explanations," whereby we explain a natural phenomenon by positing the involvement of personal agency/supernatural involvement.

Nowadays, however, scientists look for natural explanations. Rue writes, "Whereas we once explained nature in personal terms, we now explain personal reality in natural terms. The explainer has become the explained; personal agency has been naturalized; nature is now the better known reality by which we apprehend the lesser known reality of persons."[14] The efforts to ease the tension regarding the apparent conflict between science and religion (or between that of personal causation and natural causation) are, according to Rue, efforts to "mitigate the erosive effects that the scientific worldview has had on mythic realism."[15]

The second reason, Rue suggests, for the increase of nonrealistic interpretations can be traced back to the awareness of other, often competing, mythic visions. With increasing religious pluralism, it is difficult for the religious believer to maintain an exclusivist view, the idea that there is only one true religion. Rue, however, suggests that the move beyond exclusivism threatens mythic realism and therefore "the potential of mythic images to influence their emotional responses."[16] Some, having rejected exclusivism, come to adopt a pluralist view of religions, thereby maintaining that all religious traditions represent the same ultimate/transcendent reality. Rue warns that pluralism usually collapses into relativism and religious nonrealism: "But having come this far, the truncated realism of the pluralist is a mere half-step from the full-fledged nonrealism of social constructivism and subjective relativism."[17]

According to Rue, realism is not just the natural attitude with regard to religious truth talk. It is also a necessary component for religious beliefs. That is, S must believe in the objectivity of X for X being able to properly motivate S. Nonrealism, according to Rue, will undermine the relationship of motivation between S and X. We need to have an ontological commitment for religious beliefs to function in the way that they are supposed to function. Rue writes, "It is very doubtful that theistic myths, such as Judaism, Christianity, and Islam, could carry on effectively if individuals became non-realists regarding the God-as-person-metaphor."[18] He illustrates this by asking the reader what would happen if he told her that there was a tiger lurking behind her. The problem with being a nonrealist about the tiger, in this scenario, is that she "will not appraise the moment in a way that will incite fear."[19] Similarly, one being a nonrealist about God will not produce the adequate religious attitude, and so religion will not fulfill its intended function.

Donald Crosby makes a constructive and systematic case for a religious approach to nature, one that he believes is consistent with the framework of naturalism. Rue affirms a very clear realism regarding religion. Crosby's realism is less explicit. I nevertheless interpret Crosby as holding to a realist conception of religious naturalism in the way that he brings forward nature as a suitable religious object.

For Crosby, inspired by Alfred North Whitehead's process philosophy, nature is dynamic and ever changing. Nature is constituted by "restless energy of growth, nurture, productivity, and change."[20] As we saw in chapter 1, Crosby defends a religious conception of nature by considering and applying to nature six specific "role-functional categories" that are commonly associated with a religious object. The functions of a religious object are uniqueness, primacy, pervasiveness, rightness, permanence, and hiddenness.[21] By showing that it is possible to ascribe these functional categories to nature, Crosby seeks to demonstrate that nature can be recognized "as among the significant claimants for religious aspiration or devotion, alongside more familiar claimants such as God, the gods, Brahman, or Tao."[22] For Crosby, then, an objective and realist view regarding the religious significance of nature can be defended via these six role-functional categories. As will be seen, Crosby's approach is significantly different from Charley Hardwick's antirealistic and existential take on religion, and Willem Drees's instrumentalist/functional view of religion. It is also different from the pragmatic approach, which urges us to adopt a religious attitude toward nature, because such a change of attitude can help us to achieve certain pragmatic goals.

However, many would question Crosby's positive view of nature and his form of religious naturalism. A skeptical person could say, "Nature is beautiful and awe-inspiring, but is it not also true that it is wasteful, cruel, and indifferent to human flourishing? How is it possible to revere something that contains so much suffering and evil?" I will discuss Crosby's defense of religious naturalism only briefly, as it will be thoroughly analyzed in the next chapter. Crosby's response to the moral objections to a religious approach to nature is formulated in the following way: to reject religious naturalism because nature is deemed as "cruel" is to anthropomorphize nature. Hence, by applying moral and mentalistic terms to nature, by accusing it of being cruel, we are committing a category mistake. The same thing can be said, argues Crosby, regarding the objection from the indifference of nature. To say that something is indifferent is also to apply mentalistic terms, which is not possible, or even incoherent, when it comes to nature.

Crosby's further argument is that evil and suffering are necessary features of reality, that there is no such thing as a perfect world. Nature is intrinsically morally ambiguous; "reality and ambiguity go necessarily together."[23]

An imagined ideal world would lack "certain goods of our present world."[24] Freedom, as one of those goods, would not be possible in a morally perfect world, as the reality of freedom includes the capacity for good as well as the capacity for evil. In a perfect world, free from evil and suffering, we would be mere robots. Biological diversity, another good in our present world, would be absent in a perfect world, as such a phenomenon is dependent on struggles and conflicts between species, and the death and extinction of some. However, this is not to say that such occurrences in nature are goods. Crosby writes, "The sufferings, deaths, and extinctions are not goods; they are intrinsic evils. But they make the good of biological diversity possible."[25]

Another possible objection to Crosby's religious proposal is that there is no purpose in nature. Nature is simply a collection of impersonal, random processes. Meaning is an illusion in this world. Crosby grants that on naturalism, nature itself has no purpose: "It is not the outcome of conscious design, nor is it guided by the purposive will of a personal being or beings."[26] Nature exists by virtue "of its own inherent, self-contained potentialities, principles, and laws."[27] Crosby further argues that while there is no purpose *of* nature, it is possible to affirm purpose *in* nature. Naturalism still allows for teleology within the universe, the *natura naturans*, the creative aspect of nature giving rise to novel phenomena and new properties of matter.[28]

Moreover, Crosby suggests that inwardness is widespread in nature, that living organisms are not "its," but "thous." There is a "thou" character to all forms of life, meaning that a living system exhibits sentience, intentionality, and purpose. Drawing on the work of Evan Thompson, Crosby suggests that we must abandon the Cartesian separation between matter and mind with its emphasis on the primacy of the scientific third-person perspective. Indeed, the "mystery of inwardness is not confined to us humans; it reaches deeply into nature and characterizes all forms of life."[29] This leads Crosby to conclude that inwardness is a necessary feature of that life, and "the inwardness of mind, in its turn, rests upon the inwardness of life."[30] Nature as a whole is not conscious, but by virtue of the thou-character of all forms of life, and the "Thou of Nature" as a whole, it is possible to enter into an I-thou relationship with nature and nonhuman beings, despite the lack of universal purpose.[31] Nature is not a lifeless *it* but a living *thou*, thus making it possible and coherent to revere nature, and care for its well-being and flourishing.

To conclude, Crosby's defense of a religion of nature results in a realistic understanding of the religious potential of a naturalistic view of reality. It is possible, according to this approach, to adopt a religious attitude toward the natural domain, as it fulfills certain role-functional categories that are commonly associated with an object of religious devotion and reverence.

We have now explored two ways of construing religious naturalism in a realistic manner. These two proposals will be critically evaluated in the next chapter. In the next section we shall explore two antirealistic approaches, expressed by Charley Hardwick and Willem Drees.

Religious Antirealism in Hardwick and Drees

Having discussed Rue's and Crosby's realistic take on religion, we now turn to two antirealists, Charley Hardwick and Willem Drees. Hardwick agrees with Rue's sentiment that contemporary debates in theology and philosophy assume a supernaturalist view of religion, whereby an adequate conception of religion requires belief in a personal God and the idea that there is some form of final causation in the universe.[32] According to Hardwick, because of these assumptions, theological discussions are intrinsically biased against philosophical naturalism. The ontological framework of supernaturalism, however, has to be rejected, as the idea of God's personal nature, as well as the notion of God as spirit, is fraught with metaphysical problems. If Christian theology is to have a future, it needs a new ontological framework, according to naturalists like Hardwick.

Through Rudolf Bultmann's *demythologizing* project, Hardwick outlines another possibility for understanding Christian theology. The important lesson to take from Bultmann, according to Hardwick, is that a supernaturalist interpretation of Christianity and the kerygma (the message of the gospels) is not the only game in town. Bultmann's existentialist interpretative framework suggests that Christian faith is not an ontological worldview at all but an existential self-understanding, a mode of being.[33] The function of a myth, in general, and of Christian theology, in particular, is to present not an objectively true account of the world but rather a way of being.[34]

For Hardwick, an existential understanding of the Christian faith is attractive "because articulating faith by modes of existence avoids both mythology and the subject-predicate literalism of theism."[35] Indeed, for Bultmann, the kerygma is not dependent on any specific worldview. Thus this "independence of the kerygma from any *particular* intellectual preconditions is what Bultmann means by his claim that demythologizing represents the consistent application of the doctrine of justification by faith alone."[36] Bultmann, in separating the worldview of the gospels from the phenomenon of faith, provides a theoretical opening for a different ontology.

Hardwick makes the strong claim that naturalism is the default mindset for many people in today's society. When we notice the latest discoveries in microbiology, consult a doctor, or simply cook on the electric stove, we are thinking naturalistically. Hardwick suggests that today we wear naturalism almost as a second skin.[37] Hence, considering the great intellectual leap required by traditional theism, naturalism appears to be the more attractive option. A naturalistic interpretation of theology is, according to Hardwick, a wise decision, as Christian theology would, by a single stroke, "be aligned with the conceptual framework that otherwise underlies almost all contemporary culture."[38]

Given today's context, we need, in our theological reflections, to start with naturalism. The reason, argues Hardwick, for why one should adopt naturalism is that our modern world is constrained by the discoveries and methodologies of the natural sciences. To be aligned with naturalism, or physicalism as Hardwick also calls it, is to be aligned with science itself.

Given that naturalism denies any ultimate cosmic purpose, Hardwick is concerned that we will be left with a nihilist view of reality, that human existence will become absurd and meaningless. Hardwick asks, "How can anything really matter if matter does not?"[39] Is it possible to uphold a naturalistic ontology and still avoid the nihilistic conclusion? Hardwick (following John Post again) questions some assumptions regarding the view that naturalism is intrinsically hostile to meaning and purpose. It is often claimed that the absurdity of the universe automatically entails the absurdity of one's own existence. The logic of this assumption, argues Hardwick, is flawed for the following reason: to ask existential "why-questions" regarding the existential meaningfulness of physical reality is to assume that an explanation for our existence is possible. Contrary to this assumption, Hardwick maintains that "the universe is not the sort of thing to which an explanation-seeking why-question can apply."[40] In calling something absurd, or irrational, we are applying some independent standard of judgment, but such a standard is missing when it comes to issues of existential ultimacy. Hardwick writes, "We cannot show that our lives are absurd because the universe has no explanation unless we establish independently that the universe is absurd, and this we cannot do."[41] The nihilist conclusion can thus be avoided. Nevertheless, naturalism still brings with it significant changes for our understanding of religion, and the realism-versus-antirealism discussion.

We can see that Hardwick's antirealism is a product of his commitment to a physicalist ontology. To seek to ontologically ground God's existence in reality is impossible, according to Hardwick. He writes, "From a naturalistic perspective, I hold that seeking this kind of ontological foundation for religion is a dead-end. The reason is that 'God' is simply not to be found in a naturalist ontology of what exists."[42] Contrary to a metaphysical realist notion, God, in this view, should be construed "as a kind of meta-assertion that compactly expresses a valuational matrix, a form of life and an attendant seeing-as."[43] Hence, when we say that God exists, it is not to assert the existence of a supernatural being. Instead, we are effectively asserting a way of life and a way of viewing reality. This way of viewing reality, as described in chapter 1, is called a "theistic-seeing-as." The proposition "God exists" should be understood not in ontological terms but in valuational terms. The summary term for this Christian stance is "openness to the future," according to Hardwick.[44] To say "God exists" is not to make a metaphysical truth claim but to reflect a certain attitude. Therefore, "the Christian proclamation is the offer of a new self-understanding articulated in terms of events of grace. 'God exists' becomes a meta-assertion for the truth of *this* description and is a condensed summary statement for its valuational matrix."[45]

Even though Hardwick resists a metaphysically realist notion of the Christian tradition, he rejects the critique that his proposal implies theological relativism; that is, that a theological assertion is nothing more than an attitude or expression. Some form of objectivism can still be claimed within theological discourse, as the proposition "God exists" can still be cognitively true.[46] The issue facing Hardwick is to make sense of the idea of how a particular discourse about God can be true while "God" has no objective referent. It seems as if theological noncognitivism is inescapable. In response to this, Hardwick argues that theology can retain its cognitive character. A Christian seeing-as is cognitive because the values constituting this form of seeing-as are objectively true and grounded in nature. As long as there is a fact of the matter with regard to the values expressed by the Christian seeing-as, it can still be considered true in an objective sense.

Hardwick further argues that his approach is not to be confused with theological instrumentalism/pragmatism, which states that the proposition "God exists" is true not because God exists but because it is good or useful for people to have that belief.[47] Hardwick argues that his form of naturalistic theology should not be treated as pragmatic fiction, as the proposition "God exists" can still be true according to this approach, regardless of its practical value for human beings. Religious "truth can be quite independent of what we might construe as good for us or in long term best interest, as is the case with all objective truth applied to values."[48] Hence, Hardwick's physicalist approach to religion is significantly different from the pragmatic approach, which will be discussed below.

Willem B. Drees adopts a similar nonrealist position with regard to religious discourse for what seems to be two reasons. First, an evolutionary and functionalist account of religion renders an objectivist view of religious discourse very improbable. Second, religion as it is regarded on the basis of metaphysical realism clashes with human experience.

From an evolutionary perspective, one would explain the phenomenon of religion in the same way as one explains political institutions and languages. These kinds of explanations are functional, meaning that religion emerged "and therefore contributed to the inclusive fitness of the individuals or communities in which they arose, and which in turn were shaped by them."[49] The pattern of evolutionary explanations is therefore functional. Consequently, we assess an action or a process in terms of its contribution to, in this case, the inclusive fitness of the biological community. Drees's evolutionary take on religion bears a strong resemblance to Rue's approach to mythic traditions, whereby the primary function of religion is to support personal wholeness and social coherence and, thus, reproductive fitness. Contrary to Rue, however, Drees considers religious realism to be undermined by the advancements of modern science.

Drees suggests that a functional view of religious language does not *automatically* mean that the realities to which religious propositions refer have to be denied. Nevertheless, contrary to many of the concrete objects that we encounter in the world, there is a lack of "causal interaction" when it comes to religious realities and entities. Drees writes, "We refer to trees, and we seem to do so in fairly adequate ways, because our language has arisen and been tested in a world with particular ostensible trees. On the naturalist view there is no locus for particular divine activities in a similar ostensible way."[50] Hence, an evolutionary and functionalist understanding of religion seems to undermine the credibility of realist and objective religious references, since they would transcend the specific environments in which religions arose.

According to Drees, another difficulty for a realist construal of religious language concerns the clash between the metaphysics of religion and human experience. Drees writes that the realist way of understanding goes "too far beyond, if not counter to, experience."[51] Drees does not explain in what sense metaphysical assertions about divine activity go contrary to human experience. However, it is probably connected to his earlier point about the lack of causal interaction with religious entities: that it is hard to talk about divine realities in an ostensible way.

Drees explains the naturalistic interpretation of religion in terms of striving for an *ultimate ideal*, which "surpasses any actual achievable goal or situation."[52] The function of worship, given this view, is to strengthen a particular way of life and to nourish the individual and communal spirituality in relation to the "conceptions and ultimate ideals of good life."[53] This means that different conceptual elements central to many traditions, such as "'Kingdom,' 'Paradise,' 'Heaven,' 'Nirvana,' 'immortality,' [and] 'emptiness,'" function as regulative ideas against which we evaluate our behavior.[54] This kind of theological instrumentalism, as we saw earlier, was rejected by Hardwick. For Drees, a religious tradition can still be useful and powerful even if it is deemed implausible or false.[55]

In chapter 1, I discussed Drees's emphasis on limit-questions in his religious proposal. Reality is mysterious, and the more we know, the more aware we also become of the limitations of current knowledge.[56] Science enlightens our understanding of reality, but it also shows that existence is covered in mystery. Science has been tremendously successful in mapping reality and organizing knowledge. Yet through science we encounter the fundamental issues regarding physical existence. In the end, the "natural sciences may point us towards the limit-questions, but they do not get us across the boundary."[57] Even if scientists were successful in outlining a complete theory of the fundamental constituents of physical reality, "such a theory would not by itself explain why there is a reality which behaves accordingly."[58] In this way, Drees suggests that it is still possible to regard "the Universe as a gift, as grace."[59] In some ways, Drees seems to lean toward an agnostic position regarding theism and extranatural realities. This reading of science does not exclude, for example, the existence of God, but considers God's existence a genuine possibility given the mysteriousness of reality. God is, according to Drees's metaphysical understanding, "the principle of otherness."[60] Perhaps Drees does not reject religious realism but rather remains agnostic with regard to religious/supernatural entities. He also writes, "Agnosticism with respect to transcendence beyond our conceptual scheme reminds us that any proposal for an understanding of God is always limited by the conceptual scheme used."[61]

However, it seems as if Drees also maintains something stronger, which entails more than an agnostic attitude regarding God's existence: a denial of divine existence as such. Drees has argued that religious realism is not only undermined by modern science. In fact, religious realism undermines the crucial function of religion, "namely in providing a guiding vision which shapes our way of life."[62] It is in light of the problems of religious realism that Drees proposes a functional understanding of religious discourse. In the end, Drees seems to adopt an antirealist, or nonrealist, position with regard to religious truth claims. This view comes fairly close to the antirealist position advocated by Hardwick. However, a noticeable difference between Drees and Hardwick concerns the validity of limit-questions. As I described earlier, Hardwick argues that the notion of limit-questions, or "why-questions," is incoherent, as no such ultimate explanation is possible. It is a logical mistake to posit such questions. Limit-questions remain central for Drees's proposal, and it is because of the limitations of science that religion can have a place at the table, albeit in an antirealist seat.

Understanding Pragmatic Religious Realism

Although this chapter began with an outline of realistic and antirealistic understandings of religion, it still seems that a pragmatic form of religious realism is the position of the majority of religious naturalists. It will therefore be given extra attention here. How should we construe the realness of "God" or "the Sacred" on this view? I will start off by outlining pragmatism in a more general way and then move on to applying this pragmatic framework to the perspectives of Gordon Kaufman, Karl Peters, Ursula Goodenough, and Stuart Kauffman.

In following John Dewey, it can be said that pragmatism starts from the actual situation of human beings, the limitations and capacities that are involved in being a human subject. Truth, or what reality is like, is always connected to human practices, and in thinking about truth we must reflect on what practical difference the truth of a proposition is understood to make in the scheme of things. Pragmatism represents an instrumental approach to knowledge and a rejection of meaningless speculations of things unobservable, for example, abstract objects detached from human experience.[63] Characteristic of the American pragmatist tradition is the attempt to find plausible and attractive alternatives to essentialism and foundationalism.[64] The latter has been found problematic by pragmatic thinkers, since truth is available only from within a human practice, "not from any higher standpoint" (i.e., there is no God's-eye point of view).[65] A further reason for rejecting foundationalist conceptions of truth is that pragmatists are highly skeptical of commitments "to something external, whether it be reality, objectivity, or causality."[66] Richard Rorty says the following concerning the antifoundationalist strands of pragmatism: "They think that the question whether my inquiries trace a natural order of reasons or merely respond to the demands of justification prevalent in my culture is, like the question whether the physical world is found or made, one to which the answer can make *no practical difference*."[67] Essentialism is rejected for quite similar reasons. There is no unconceptualized ready-made world out there waiting to be discovered; there is no such thing as uncontextualized truth. Hence, what we deem as "real must be relevant to our *practical interests*."[68] For example, a pragmatic conception of scientific theories suggests that we should prefer scientific theory X over Y, not because X more successfully describes some part of an external and unconceptualized reality, but because X enhances our ability to control and predict events taking place in the world. That is, the merit of a scientific theory is connected to its practical advantages.

The world is, then, in a sense a pragmatic construction. However, it is primarily an *ethical* or a *moral* construction, since we are responsible for what we do in the world and how we treat other people, and this is always related to ontology.[69] Thus the truth of a proposition should not be judged in terms of whether it successfully or unsuccessfully refers to a mind-independent reality; rather, truth is the practical result of inquiry, "and society informed by science is the best judge of what works."[70]

Roughly speaking, a pragmatic philosophical approach to truth, objectivity, and human practice seems to involve the following ideas:

1. We must start all philosophizing by conceding the naturalness of human beings and that we are embedded in human practices.

2. Truth must make some practical difference.

3. Truth is accessible only from within human practices.

4. Our ontological commitments and what we consider as true always involve an ethical dimension (there is no ethically neutral epistemological or ontological position).

5. Truth is the practical result of inquiry, and society (informed by the best science) determines what works. Hence, society determines what is true and what is not.

It is not a stretch to say that pragmatism takes a functionalist approach to truth assessment. Nicholas Rescher writes, "Its root idea is that virtually every human enterprise and endeavour has an aim—a purpose or goal of some sort—and that the natural test of adequacy is afforded by the efficiency or effectiveness in meeting this objective."[71] The functional efficacy of a certain human practice is the defining objective for pragmatism. However, this form of pragmatism is also, in a qualified sense, *realist*, "because it is realist about the environment in which we actually live, the practice-constituted reality in which talk of true and false, real and unreal, has its proper home."[72] Hence, contrary to the antirealist rejection of truth, the pragmatist accommodates the concepts of "truth," "false," "real," and "unreal," but these concepts must always be judged with regard to human practices and needs.

Pragmatic Realism Applied to Religious Naturalism

Let us then consider what pragmatic realism means for religion and its truth value. Pragmatists ask us to think of truth in terms of practical results, what is good for us to believe and what difference it makes for our socially shared human practices. Applied to religion, one might say that religious truth is not about certain propositions corresponding to a religious reality, for example God, but rather that they enable people to live a good life in the social community in which they find themselves. The truth of religious beliefs is thereby relative in that they depend on the practical interests of people; and given that human practices may change over time, so can religious truth.

We may find out that certain religious beliefs held by people have produced disastrous consequences, such as in cases of religious fundamentalism and religious extremism, and we should therefore revise them or reject them completely. A pragmatic conception of religion, furthermore, involves the rejection of religious metaphysical realism. A pragmatist can still talk about some religious beliefs being true and some being false, but their truth value is not to be decided by means of seeing if certain propositions correspond to some independent reality, because reality and truth are accessible only from within human practices. The pragmatist criterion for religious (PCR) truth is therefore:

> The truth of religious belief X should be judged in terms of the practical consequences of adopting X.

It is not obvious if PCR applies only to individual persons or to social communities as well. That is, when one considers the practical consequences of belief *X*, should one judge them only with respect to what they might imply for individual persons, or should we broaden the pragmatic consideration as to include social communities? If so, how wide should we cast the net?

As mentioned earlier, several ecoinclined religious naturalists hold two aspects to be of great import when evaluating religious truth claims and whether we should reject or embrace a given religious tradition: first, what consequences religious belief *X* might carry with respect to the health of the ecosystem; and second, what religious belief *X* might mean for our religious/spiritual life. In chapter 1, I argued that several religious naturalists stress the importance of adopting what I have termed an "EMA," an ecologically mindful attitude. We are being called to adopt attitudes that serve the health of the ecosystem and make us more likely to act in an ecologically responsible manner.

Some religious naturalists call for a change to the semantic content of religion in order to better produce EMAs. As Karl Peters puts it, we need new metaphors, and we "need to see ourselves as webs of cosmos, life, and culture, so that we and the rest of our planet can continue and flourish."[73] Gordon Kaufman argued that the constructive task of theology is to reconsider some long-held religious beliefs (such as a dualistically construed conception of God) in light of the ecological crisis "that so urgently demand[s] our attention."[74] Basically, this means that the theologian's task is to distinguish between those beliefs that promote an EMA and those that do not. The truth of religious propositions is therefore judged in relation to us adopting an EMA. The notion of EMA, in conjunction with PCR, implies the following criteria for religious truth claims (PCR^{EMA}):

> The truth of religious belief *X* should be judged in terms of the practical consequences of adopting *X*, and the practical consequences of *X* should be judged in terms of us becoming more likely to adopt an EMA.

We have seen above how a pragmatic conception of religion applies to religious naturalists in a general way, particularly with regard to the development of EMAs. I will now provide a more thorough description of the religious proposals of Gordon Kaufman, Karl Peters, Stuart Kauffman, and Ursula Goodenough, and how each perspective relates to pragmatism.

Pragmatic Dimension of Gordon Kaufman's Proposal

Gordon Kaufman explored and wrote extensively on the concept of "God" during his theological career. He has not always been considered a religious naturalist but moved in the direction of a more robust naturalistic theology in his later work. It can be argued that the major guiding point for Kaufman's earlier theological ambitions was the skeptical philosophy of Immanuel Kant. Later, however, while not abandoning Kant completely, Kaufman came to prioritize the evolutionary narrative for understanding the function of the symbol of "God" in a scientific age. It is also during this time, when Kaufman seriously started to consider the impact of ecology on theology, that a pragmatic dimension started to emerge in his theology.

I will start by outlining Kaufman's earlier reliance on Kant. What Kant showed, according to Kaufman, was the impossibility of direct and immediate experience of reality.[75] This idea changes the game for theology dramatically. Kaufman adds that this Kantian idea has been tacitly assumed in theology for a long time. The Christian tradition has historically affirmed that God is "not an object of ordinary perception" but is transcendent and essentially mysterious.[76] This inaccessibility of God is a frequently repeated theological theme found in, for example, the biblical story of Job.[77]

When others speak of *experiencing* God, Kaufman confesses that he does not understand what it means. By following Kant further, Kaufman writes that he eventually came to conclude that "talk about *experience* of God involves what philosophers call a 'category mistake.' "[78] God does not exist as an independent person, being, or reality, and so cannot be directly experienced. Instead, Kant shows that the concept of God can be understood only as an imaginative construction of the mind.[79]

Kaufman, in adopting Kant's epistemology, makes the distinction between the "available referent" and the "real referent." The former refers to "a particular imaginative construct—that bears significantly on human life."[80] It is this available God whom we have in mind in worship and in prayer. However, the real referent, God, is not available to us due to our cognitive limitations and an unavoidable epistemological filter, and will "always remain a transcendent unknown, a mere point of reference."[81]

It is in his book *In the Beginning . . . Creativity* that Kaufman takes a defining step in the direction of naturalism. Kant, while still important for Kaufman, is no longer forming the theoretical background for Kaufman's theology. Instead, the evolutionary narrative sets the boundaries for theological constructions, with regard to the concept of God and Christology.

A guiding principle for Kaufman is that human life should be understood from a *biohistorical* perspective. This perspective, Kaufman states, "emphasizes our deep embeddedness in the web of life on the planet Earth while simultaneously attending to the significance of our radical distinctiveness as a form of life."[82] The biohistorical framework takes into account the long evolutionary history of human nature. However, this perspective further emphasizes those aspects of human nature that seem to go beyond, and indeed cannot be explained solely in terms of, biological language. That is, "reductionistic physico-biological naturalisms will not enable us to understand ourselves adequately or to fit ourselves appropriately into our ecological niche on planet Earth."[83] In order to understand the nature of humanity, our distinctive characteristics, and our place in reality, we need to pay attention to the historical aspect of human evolution.[84]

In the end, humanity can be fully understood only if we harmonize the biological and the historical aspects of human evolution. The *biohistorical* view is an attempt to avoid a one-sided approach to the complex issue of human origin. Of course, that humans emerged at all is a result of biological evolution. Nevertheless, what makes humans distinctive "came about through the growth of historic-cultural processes that helped to push the development of *Homo sapiens* in decisively important directions."[85] Thus Kaufman, drawing on anthropologist Clifford Geertz, suggests that the human family is both a cultural and biological product.[86]

This biohistorical understanding proves helpful for more than simply explaining the phenomenon of humanity within an evolutionary framework. For Kaufman, such emerging understanding can shed light on the function of "God," one of humanity's most important symbols, in a scientific and ecological age. A biohistorical understanding of humanity breaks with the anthropocentric tendencies of Western religious traditions. Humans, it has been assumed, were created in the image of God, and this human-centered and dualistic theology made humanity into the climax of creation. According to Kaufman, such a view is untenable given the strong embeddedness of humans within the evolutionary web of life, plus the strong physicality of human beings. "God" functioned during most of Western history as the ultimate point of reference through which all human life should be understood. In today's context, by contrast, we should start from nature in order to understand both God and humanity.

The symbol of God is used to remind ourselves to participate in the mystery that is continuously being manifested in the world.[87] "God" signifies the novel and mysterious phenomena that are brought about by the evolutionary process. This form of mystery is also referred to as "serendipitous creativity," suggesting that the universe is not static or a "permanent structure, but rather as constituted by (a) ongoing cosmic serendipitous creativity that (b) manifests itself through trajectories of various sorts working themselves out in longer and shorter stretches of time."[88] A biohistorical understanding of reality, which gives emphasis to the creativity of the universe, profoundly changes our conception of and relationship with God.

It is in the light of the current ecological crisis that we need to actively reconstrue God as the creative process, the mysterious unfolding of nature. In this way, Kaufman's pragmatic reorientation of religion focuses on developing images of God that are conducive for motivating people to care for the ecological well-being of planet Earth. All that I have described above is motivated by this ecological concern. Kaufman retains, however, a form of realism (as was also discussed in chapter 1), as God is identified with an objectively real feature of reality, that is, the evolutionary processes that give rise to novel phenomena. Religious truth, in this pragmatic view, is formulated through and judged against the criteria of PCR^{EMA}. Hence, Kaufman's theological ambitions result in a theological ecopragmatism, whereby religious images are rethought and judged in terms of their ecological adequacy. As we will see in the next section, Karl Peters, who draws insight from Kaufman's theology, proposes something similar.

Pragmatic Thinking in the Theology of Karl Peters

In a similar way to Kaufman, Karl Peters is committed to transforming the Christian tradition from within by utilizing naturalistic insights regarding nature, ecology, and religion. Traditional religions, argues Peters, offer insights into the sacred dimension of reality. Yet they ultimately fall short when it comes to offering a *practical theology*, a theology that one can live by.[89] The aim of Peters's theological project is to offer a new image of God, or the Sacred, by which one can meaningfully live. Like Kaufman, Peters suggests that our images of God must be consistent with the epic of evolution. This has been one of the major problems for traditional conceptions of God. The religious naturalist is "looking for a more integrated understanding of God and evolutionary theory, of God working creatively in the world and of Darwin's theory of random variation and natural selection."[90]

Peters follows the tradition of empirical theology that claims that the adequacy of religious beliefs must be tested against human experience. Hence, as supernaturalistic versions of God—by virtue of positing God as a supreme, transcendent, personal being—do not fall within the purview of human experience, they must be rejected. The empirical theologian Ralph Wendell Burhoe exerts a major influence on Peters's thinking. Peters identifies four ways in which Burhoe has contributed to his own theology. First, he offered a way of thinking in relational terms. Second, Burhoe provided an understanding of religion that made it possible to appreciate religious phenomena on scientific grounds. Third, he offered an evolutionary understanding of how religion adapts and changes. Fourth, Burhoe showed how God can be "understood in terms of function and system."[91] Having rejected traditional religion, Peters found with the help of Burhoe a way to "formulate theology that was in accord with scientific theories, scientific methods, and a scientific view of the world."[92]

What notion of God is possible, from Peters's perspective, within the emerging scientific worldview? As mentioned above, Burhoe enabled Peters to develop a view of God in terms of function and system. Peters, employing Kaufman's idea of serendipitous creativity, suggests that this concept "points to a system, the parts of which work together in unpredictable ways to create such things as new life, new truth, and new community."[93] Furthermore, Peters suggests that Kaufman's notion can be used to talk about the "religious significance of biological evolution."[94] Peters proposes that with this idea "of God as the universal creative process, continuously at work to give rise to new forms of existence," we can gain a deeper appreciation of our place in the cosmos.[95]

Peters follows Kaufman in adopting a pragmatic perspective on theology.[96] Based specifically on Charles Sanders Peirce's pragmatism, Peters's view suggests that God can be "conceived in terms of how humanity relates to—acts on and experiences—this same thing."[97] A pragmatic way of justifying theological concepts means that we judge "the concept's effectiveness in shaping behaviors that help people to ever-greater richness of experienced value."[98] This is a general form of pragmatism, according to Peters. However, we have also seen that, in placing the function of religion in relation to the well-being of the ecosystem, Peters turns ecological stability and flourishing into a pragmatic goal. This pragmatic goal in turn shapes and regulates our conceptions of God. By placing God firmly within the natural order and viewing God as the process of creation itself, Peters claims, we are more likely to work actively to support the integrity of nature. Peters goes on to remark, "[In] trying to live in harmony with the divine creativity, I feel morally responsible for how I live. Once I recognize my kinship with all of life, I realize my human responsibility to care for other members of our natural family."[99] Peters therefore adopts a pragmatic approach to religion, and argues that theological constructions should be judged according to the likelihood of them inspiring us to develop a harmonious relationship with nature. We can see, then, the ways in which Peters clearly affirms PCR^{EMA} as the overarching logic for understanding theology.

Stuart Kauffman's Reinvention of the Sacred

Whereas Kaufman and Peters seek to transform a religious tradition from within, others start off with naturalism and then incorporate particular beliefs from traditional religions. Stuart Kauffman, as we saw in chapter 1, represents the view that I call "religiously informed naturalism." In order to enable a religious or spiritual interpretation of science, Kauffman addresses the problem of reductionism. That there is a reductionist view of science and that such a view is widespread is not surprising: "We seek reductionist explanations. Economic and social phenomena are to be explained in terms of human behavior. In turn, that behavior is to be explained in terms of biological processes, which are in turn to be explained by chemical processes, and they in turn by physical ones."[100] It seems as if methodological reductionism is already built into science and that a reductionist ontology follows naturally. Consequently, we are left with a worldview according to which "all of reality is *nothing but* whatever is 'down there' at the current base of physics."[101] This reductionism turns all the aspects of the universe that we take to be sacred—agency, meaning, values, purpose, and so on—into mere illusions. In concluding that we live in a meaningless universe, a naturalistic view of reality becomes the antithesis of a religious conception of reality.[102]

We saw in chapter 2, where I explicated the naturalistic dimension of religious naturalism, that Kauffman subscribes to a thoroughly pluralistic ontology: reality is layered, emergent properties exist in a realist fashion, and meaning, values, and consciousness are real phenomena. A large part of Kauffman's project therefore involves a stringent critique of reductionism and the idea that basic physics defines all of reality. I will not repeat this critique. Instead I will focus on Kauffman's understanding of the function of religion.

Given the failure of reductionism, Kauffman maintains that the biosphere is essentially unpredictable, and that "mystery" is real. Yet this mystery does not entail the need for a supernatural God, as nature itself "has given rise to all that we have called upon a transcendental Creator God to author."[103] The main motivation for Kauffman's ambition to "reinvent the Sacred," to sacralize the natural domain, is the need for a global ethic. This ethic must "embrace diverse cultures, civilizations, and traditions that span the globe."[104]

Moreover, such a global ethic must be of "our own construction and choosing."[105] Kauffman warns us that without such an ethic we will head for a global ecological disaster. For Kauffman, however, a global ethic will not provide all the ethical guidelines that we might need and desire. On the contrary, a global ethic will encourage humility in light of the unpredictability of the biosphere. Indeed, any "global ethics we create must embrace our inevitable ignorance as we live our lives forward" into mystery.[106]

In a similar way to Kaufman and Peters, Kauffman stresses the need for us to actively participate in the Sacred unfolding of the universe.[107] However, as Kauffman notes, awe and respect have become unfashionable in our postmodern society.[108] This is where religious symbols become relevant for Kauffman's project. A global ethic needs to be anchored in something that spans many cultures. In the same way that it is our responsibility to formulate a global ethic, it "becomes our choice what we will hold sacred in the universe and its becomings."[109] The mysterious character of reality is ontologically real, on Kauffman's approach, but how we conceptualize the mystery of the biosphere from a religious perspective boils down to choice.

Thus Kauffman's "religiously informed naturalism" is pragmatic in nature. The quest to retrieve the symbol of "God" is due to the need to "find a global spiritual space that we can share across our diverse civilizations."[110] In response to a metaphysical realist understanding of religion and God, Kauffman argues that a reinvention of God/the Sacred is possible given that "God" is our own invented symbol. He writes, "It is we who told our gods and God what is sacred, and our gods or God have then told us what is sacred. It has always been us, down to millennia, talking to ourselves."[111] Given the current global crisis, Kauffman urges us to talk to ourselves consciously, and actively choose our own ecologically mindful conception of the Sacred.[112]

Ursula Goodenough's Modest Pragmatism

Ursula Goodenough seems to express a pragmatic understanding of the nature and aim of religious naturalism. Yet, as we shall soon see, Goodenough's version is different in many ways from the pragmatism that has been discussed so far. Like Stuart Kauffman, Goodenough's push for religious naturalism is ethically motivated, and it takes into account our current ecological situation. We need on this view (as was discussed in chapter 1) a planetary ethic. That such an ethic is necessary becomes obvious, argues Goodenough, when we consider climate change, nuclear weapons, and pollution.[113]

Humanity is facing a global crisis that can be countered only with a global solution. This is where religious naturalism comes in. A religiously naturalistic vision of reality, in relying on science, can provide the necessary culture-independent framework for a global ethic. She argues that "the story of Nature has the potential to serve as the cosmos for the global ethos that we need to articulate."[114]

But how does Goodenough view religion, and what is its function? A central term for her project is *religiopoiesis*. Concerning this term, she writes, "The poiesis part of religiopoiesis comes from the Greek *poiein*, to make or craft, the same root as poetry. Religiopoiesis, then, is the crafting of religion."[115] Anthropology teaches us that we humans are unavoidably religious. In this regard, humanity is truly unique. Indeed, argues Goodenough, humans might as well be called "homo religiosus."[116] Our clear need to find answers to the deepest existential questions is a part of our nature. Goodenough considers religious naturalism, as primarily expressed in her book *The Sacred Depths of Nature*, to be a contribution to the global project of *religiopoiesis*.[117]

On Goodenough's view, the function of religion is to "integrate the Cosmology and the Morality, to render the cosmological narrative so rich and compelling that it elicits our allegiance and our commitments to its emergent moral understandings."[118] Moreover, the core of a religious orientation encompasses three dimensions of human experience. The first dimension is the *interpretive dimension*, that is, human responses to the "Big Questions," such as "Why is there anything at all?," "Does the universe have a purpose?," and "Why is there evil and suffering?"[119] The second dimension is the *spiritual dimension* of human experience, which refers to responses to existence in the form of gratitude, awe, humility, and reverence. The third dimension is the *moral dimension*, which describes our outward responses, such as care, compassion, trust, and commitment.[120] Goodenough further argues that an emergentist-based form of religious naturalism can positively account for all three religious dimensions of human experience.

Religious naturalism provides an interpretive framework through an emergentist account of evolution: the biological community on earth, including human beings, is not brought about by a supernatural being.[121] All life on Earth shares a common ancestor, and is the result of a multitude of natural processes and specific environmental conditions. Hence, "all the creatures on the planet share a huge number of genetic ideas."[122] Science tells a story not just about dead static matter but also about a "process of becoming."[123] Reality escapes any simplistic reductionism, and science does not land us in a nihilistic view of reality.

Goodenough is gravely concerned about the nihilistic implications of naturalism but claims that she has found a way to defeat it. She writes, "I have come to understand that I can deflect the apparent pointlessness of it all by realizing that I don't have to seek a point. In any of it. Instead, I can see it as the locus of mystery."[124] By facing the ultimate questions of physical existence, we can, argues Goodenough, enter into a "covenant with mystery."[125] Such questions can be openly encountered without one trying to provide definitive answers.

Religious naturalism offers a well-needed space for spirituality: to contemplate the workings of nature is to be "invaded by immanence."[126] Nature is not a value-free domain but calls us to actively engage with it. Goodenough writes that nature calls us "to marvel at its fecundity. It also calls us to stand before its presence with deep, abiding humility. Earlier we sanctified the self, and soon we will consider ways to think about our humaneness with reverence and pride."[127] A spiritual relationship with nature, in religious naturalism, is not just possible; mindful reverence is at the heart of religious naturalism, and it is a defining feature of Goodenough's *religiopoiesis*.[128] Mindfulness can even be considered to be the foundation of religious naturalism and what it is all about. Goodenough writes, "We can take religious naturalism, translate *religious* as 'reverent' and *naturalism* as 'mindfulness,' and recognize that the orientation we are calling religious naturalism can be said to be embedded, at core, in reverent mindfulness."[129]

The religiously inspired naturalist, argues Goodenough, is called to be mindful about "our place in the scheme of things," "that life evolved, that humans are primates," of "the fragility of life and its ecosystems," "that life and the planet are wildly improbable," and "that all of life is interconnected."[130] The point of this mindfulness practice is to bring us into a state of mindful caring for nature and its creatures. Indeed, Goodenough states that the function of mindfulness in the context of religious naturalism is to produce a "mindful virtue" within us.[131]

Religious naturalism offers a naturalistically grounded eco-ethics: For Goodenough, the spiritual practice of mindfulness is intimately connected with eco-ethics. A mindful reverence of the workings of nature generates awe, wonder, and respect for the integrity of the ecosystem. A planetary ethic and an ecomorality flow from being a practicing religious naturalist. Goodenough adopts a practical approach to mindfulness, suggesting that it can help us to reconfigure our social emotions and "enlarge our moral vision such that we can come to care not just about family and troop and tribe but about conserving ecosystems and sustaining biodiversity."[132] Hence, we are not practicing mindfulness simply for the sake of a pleasant state of being that such practice can lead to but ultimately to reconfigure our ethical system so as to widen our moral horizon.

I suggest that the ecomorality and planetary ethic espoused by Goodenough takes us back to the issue of pragmatism in religion. Goodenough's proposal fits, to an extent, within a pragmatist framework. She recognizes the global crisis facing humanity, for which we must find global solutions. The function of religion, in this context, is to motivate, emotionally educate, and enable us to respond to the current ecological situation. In a similar way to pragmatic thinkers, Goodenough claims that we need to actively participate in the crafting of religion. Goodenough's proposal regarding *religiopoiesis* comes close to what Kaufman calls the "imaginative task of theology." *Religiopoiesis* is, as we have seen, not just a fact of human existence but something with which Goodenough urges us to participate. Due to this proposal's eco-ethical orientation, the well-being of nature and the flourishing of the ecosystem are construed as a primary pragmatic goal. Thus, in light of this pragmatic goal, Goodenough sets out to construct a religious worldview, utilizing the major empirical theories of science and the evolutionary narrative in particular.

However, there are still significant differences between Goodenough's approach and the pragmatic approach that is expressed by Peters, Kaufman, and Kauffman. Overall, Goodenough is less confrontational when it comes to traditional religion. Traditional religions, such as the monotheistic traditions, offer profound wisdom, and Goodenough does not seek to replace them. Indeed, Goodenough's proposal makes "no claim to supplant existing traditions but would seek to coexist with them."[133] In this way, Goodenough, contrary to the other pragmatic naturalists, does not call for a radical revision of traditional religions but opts for peaceful coexistence.

Conclusions

This chapter has explored realistic, antirealistic, and pragmatic understandings of religious naturalism. For Loyal Rue, religion cannot fulfill its intended function unless it is based firmly on a realistic framework, which is to enhance personal wholeness and social coherence. Crosby, in contrast, provides a more positive account of religious naturalism. In his view, nature, by virtue of fulfilling the role-functional criteria, can be properly called a religious object.

Charley Hardwick's and Willem Drees's antirealist accounts of religion were also discussed. Hardwick, starting from a physicalist view of reality, employs an existentialist framework for understanding the Christian tradition. Hardwick argues that religious utterances should be taken not realistically but as meta-assertions for a way of being in the world. Drees, quite similar to Rue, adopts an evolutionary view of religion. However, Drees, contrary to Rue, views religion in an antirealist manner. For Drees, religious realism undermines the function of religion and should therefore be rejected.

In the last part of this chapter I discussed a pragmatic understanding of religion. Proponents of this view include Karl Peters, Gordon Kaufman, Stuart Kauffman, and, to some extent, Ursula Goodenough. These naturalists argue that our understanding of God needs to change in light of science in general, and our ecological situation in particular. "God," if we choose to retain this powerful symbol, should be imagined as the immanent Creativity of the universe. The function of religion, according to this form of pragmatism, is not to provide us with mind-independent truths regarding a supernatural reality. Instead, the function of religion is to inspire us to adopt attitudes beneficial to the well-being of nature.

5

The Problem of Religious Discourse in Religious Naturalism

Religious naturalists offer, as we have seen, three different ways of understanding religious discourse: the antirealist perspective proposed by Charley Hardwick and Willem Drees, the realist perspective argued for by Loyal Rue and Donald Crosby, and the pragmatic realist perspective that is expressed by Gordon Kaufman, Stuart Kauffman, Karl Peters, and Ursula Goodenough. These three accounts differ substantially on the function of religion and the nature of religious truth claims. The main point of this chapter is to bring to light some of the challenges facing these ways of comprehending and approaching religion from a naturalist point of view. The final section of this chapter will be dedicated to exploring the problem of evil and suffering in relation to Donald Crosby's religion of nature.

The Tension between Physicalism and Christian Faith

Hardwick offers an interesting reinterpretation of religion, and of the Christian tradition in particular, in light of a physicalist metaphysics. Hardwick's ambition is not to *prove* a physicalist reading of Christianity. Instead, given the problems that he sees with a personal and dualistic conception of God, Hardwick argues that we must search elsewhere for an adequate ontology that is compatible with a scientific worldview. Hardwick subscribes to what I call a monistic ontology and epistemology (as outlined in chapter 2). Such monism claims that the natural world is all there is, that whatever exists is either material or a property of something material, that all truth is determined by physical truth, and that physics offers the best description of reality (*MON1–4, MEN1–2*).

Through physicalism Hardwick seeks to offer a *moral* or *ethical* reinterpretation of Christianity. Although interesting and thought-provoking, I see some difficulties with regard to the physicalized Christianity offered by Hardwick. In order to demonstrate the validity of a physicalist Christianity, Hardwick needs to show how physicalism can allow for the moral values involved in a Christian seeing-as, and how Christianity is the *best* (or most correct) seeing-as. It will be seen that his existentialist interpretation of Christianity is not grounded in physicalism, which seems to suggest that there is a tension between these two aspects of his project.

Hardwick, instead of offering an account of how we can derive a Christian seeing-as, admits that physicalism and a formal ontology of human existence "does not dictate any ontologically grounded way of being."[1] Furthermore, he posits, "the Christian faith is a genuine 'taking' [an act of faith], not something that is grounded in or otherwise legitimated by a general account of being independently of this taking."[2] That is, the physicalist framework does not necessitate any specific way of being, or any particular seeing-as or taking-as. However, on Hardwick's view, "all truth is determined by physical truth" (*MEN2*), and so the normative truth of a Christian seeing-as/taking-as *should*, following the logic of his own project, be derivable from the basic truths of the physical world. It is striking, then, that Hardwick does not ground normativity in physicalism; instead he suggests that the "valuational claim Christianity makes *is* normative."[3] Hardwick gives up on this explanatory task. He refrains from explaining how the normative truth of Christianity can be derived from physical truths, or how the values particular to a Christian seeing-as can be grounded in a physicalist ontology. He instead grounds normativity in the Christian tradition itself. Hence, the justification of a Christian seeing-as is made independently of the ontological claims of physicalism regarding the structure of reality.

It seems as if Hardwick, because he's stressing the uniqueness and particularity of Christian discourse, is hesitant about making the Christian tradition subordinate to some other framework, including physicalism. However, this creates a dualism between the religious aspect and the overall ontology of Hardwick's project, which subsequently undermines the prospects of a unified Christian physicalism.

The tension between a Christian understanding of reality and physicalism becomes even more obvious when one considers some of the thinkers that have influenced Hardwick's theology. In explaining the idea of "being in the world" (or taking-as), Hardwick turns to Martin Heidegger. This influential German thinker took religion to be a modification of a general existential structure of human possibilities. Religion is a way of existing. At first glance, Heidegger's phenomenological and existential interpretation seems to go quite well with, and indeed offers some ontological resources for, Hardwick's theological project. However, as Hardwick himself points out, Heidegger was an *antinaturalist*.[4] Hardwick, though, argues that the phenomenology of Heidegger bears a strong resemblance to physicalism: much like the physicalist, the phenomenologist stresses that one's way of existing is determined by "ontological unifiers."

It should be pointed out that Hardwick's understanding of these "ontological unifiers" differs substantially from that of Heidegger's. For Hardwick, the ultimate unities can be captured by the sciences, especially physics, and all truths are determined by physical truths. Heidegger, in contrast, argued that the different sciences, such as physics, chemistry, and biology, are unable to reveal the fundamental conditions of human existence (although he recognized their important contributions). There is a limit to naturalism, which is recognized by Heidegger, that goes contrary to the ambition of Hardwick's physicalism. It remains unclear, then, to what extent Heidegger's phenomenology can aid Hardwick in construing a naturalistic-existentialist reading of religion.

A second concern regarding Hardwick's physicalist understanding of Christianity relates to the grounding of values in natural facts. For Hardwick, as we have seen, values are central to a Christian seeing-as. A Christian seeing-as is an existentially grounded valuational stance and a way of being in this world. Hardwick is aware that he needs to preserve the normative dimension within his physicalism, and show how physical determination does not entail a negation of the moral domain. Hardwick suggests that John Post's physicalism is able to do just that.

Post argues that values are part of the fabric of the world, and they are "part of the fabric in the sense that the correctness of a value judgement is determined by purely natural facts."[5] Determination, in this context, and which I touched on in chapter 3, means that "given the way the first is, there is one and only one way the second can be."[6] Hence, given a set of natural properties, the values instantiated will be the same in $world_1$ as in $world_2$. Post summarizes this determinative relationship in the following way: "The world determines moral truth in *P*-worlds *iff* [if and only if] given any *P*-worlds W_1 and W_2 in which the entities have the same natural properties, then the same moral judgements are true in W_1 and W_2."[7] Post's aim is to protect the coherency of physicalism by bringing facts and values closer to each other, and so avoiding a potential ontological split. He is also honest about the determinist implications of holding to a supervenience relationship between values and natural properties. In holding to supervenience one is saying that moral properties "supervene on natural properties in the sense that nothing can differ in its moral properties without differing in its natural properties."[8] The physical level ontologically dictates the distribution of values. However, is this not to reduce the moral realm to the realm of physics and the domain of basic physical properties? What happens to the autonomy of Hardwick's Christian seeing-as if the values particular to that seeing-as are ontologically determined by natural properties? Hardwick seeks to offer an objective account of value, and construe religion naturalistically, while preserving God-talk within naturalism.[9]

Nevertheless, I think that Post's, and thus Hardwick's, defense of the autonomy and objectivity of values is too weak. Post/Hardwick argues that even if we had access to the full list of all natural facts and their interrelations, we would still not be able to "tell which moral judgements are true."[10] Knowing the relevant natural facts does not mean that we can derive or "justify any of the moral truths from descriptive truths."[11] Despite the supervenience relationship between facts and values, our knowledge of physical facts does not provide us with a complete list of moral truths or ethical obligations. Post/Hardwick further maintains that values have an explanatory role to play within naturalism. Value judgments are irreducible explanations. Hardwick writes, "Nevertheless, it seems increasingly clear that positing such correctness does play an important part and ineliminable explanatory role and is therefore explanatorily intelligible."[12] This is the case, since moral properties are irreducible to the physical level, in the sense that we cannot deduce moral properties from natural properties. Hardwick concludes that "the autonomy of morals therefore is secure, even if moral truth is determined by purely physical descriptive truth."[13]

We can see how Post/Hardwick launches an epistemological defense of values, stating that we have no way of deducing values from purely physical facts. The problem is that *epistemological* ignorance does not secure the positive *ontological* status of values and moral properties. In a similar way to Hardwick's and Post's defense of semantic normativity (as seen in chapter 3), they seem to suggest that the epistemological irreducibility of a higher-level phenomenon safeguards it from ontological reductionism. This argument does not hold up. It might be the case that values are ontologically reducible to natural properties, even if it is not possible to derive values from facts. Indeed, given that Post and Hardwick maintain that "everything is physical," it seems that they cannot uphold the ontological distinctiveness of values. The physical determination by natural facts implies a reduction of the moral realm. However, if this is the case, then Hardwick's construal of Christian physicalism leads to a significant conflict between the ontological commitments of his proposal.

We should conclude that Hardwick's valuation theism, and the particular values associated with a Christian seeing-as, is not safe from a physical reduction. It becomes very hard to maintain the realness of values if we also hold to physicalism and a determinative relationship between the domain of values and natural processes. Perhaps this is why, as we saw earlier in this section, Hardwick does not explicitly ground Christian discourse in physicalism, as values do not seem to fit within this framework. Hardwick, it seems, is aware of a potential metaphysical grounding problem regarding values within a physicalist ontology.

It should also be noted that the problem for Hardwick's physicalism is not just to create space for values but also to show how the Christian seeing-as, among all the different forms of seeing-as, is the correct one; that given how the world is, this is the correct distribution of values. Yet, as was pointed out earlier, Hardwick admits that Christian discourse cannot be derived from physicalism. As it was said, Christian faith is a genuine "taking," not something that is philosophically grounded "in an independent ontology."[14]

Instead of seeking to assess the correctness of a Christian seeing-as through physicalism, Hardwick suggests that it is possible to evaluate religious modes of beings in a "cross-cultural perspective."[15] This means that religious traditions will be assessed in terms of how they enable authentic existence and openness toward the future. The task, according to this approach, is to compare and evaluate different existential articulations of the human condition.

It seems here as if Hardwick is assuming a guiding value, namely "openness to the future" (which coincidently is a core feature of a theistic/Christian seeing-as), but leaves this value unexplained. Once again, a core feature of Hardwick's theology floats free of his physicalist ontology.

To conclude, Hardwick's existential interpretation of Christianity is not properly grounded in, and is even in tension with, his overall physicalist ontology. As a matter of fact, Hardwick seems to be aware of the problem of grounding a Christian mode of being in, or deriving a particular mode of being from, physicalism.

Limit-Questions and the Status of Naturalism

The naturalism of Willem Drees is different from Hardwick's/Post's physicalism, as it recognizes the intrinsic limitations of science: that science cannot, and will never be able to, offer us adequate answers to some of the deepest metaphysical questions pertaining to human existence. The strength of Drees's position, and what effectively sets him apart from naïve scientism, is that it is humbler with regard to the reach of empirical enquiry. Because of these seeming scientific limitations, an opening is created for a religious worldview. In this way, and contrary to Hardwick's globalized physicalism (that physicalism is true with regard to the totality of existence), Drees's form of religious naturalism can communicate with certain forms of noninterventionist and nontemporal theisms. Consequently, this version does not automatically invite atheism. In the end, this is a much more robust religious position, especially when compared to the physicalist construal of religious naturalism offered by Hardwick, according to which there is nothing religiously significant about physical reality. Hardwick and Drees, as shown in chapter 2, both share a commitment to monistic naturalism. Yet Drees, who places some restrictions on science and thus avoids an antimetaphysical position, can more effectively incorporate religious elements.

Drees's emphasis on limit-questions, however, raises some concerns for his naturalistic framework. If it is the case that science is unable to address the deep metaphysical questions pertaining to the nature and structure of reality, if science is intrinsically limited to the study of natural processes and phenomena, then how can Drees confidently maintain that his naturalism is superior to alternative and nonnaturalistic ontologies? According to Drees, science will never be able to rid existence entirely of mystery. As he writes, "I believe that science will not remove this 'horizon of not knowing.' There will always be 'a mist where our questions fade, and no echo returns.'"[16] Although fundamental physics and cosmology help us to probe deeper into reality, these sciences also reveal those limit-questions "which arise at the speculative boundary" of our empirical quest.[17] These questions are not deemed meaningless by Drees, but are celebrated as integral to his religious worldview. As suggested, it is the persistence of such limit-questions that sets Drees apart from those voices that suggest that this world is ultimately devoid of purpose and meaning. Drees expresses clearly an agnostic spirit in embracing those questions that seem to lie beyond the reach and competence of the natural sciences.

Nevertheless, Drees is a committed naturalist and maintains that the natural world is a unity in the sense that all entities, properties, and objects are made up of the same basic constituents. It is because of this "constitutive reductionism" that Drees rules out dualism, as well as supernatural interventions in the natural order. As Drees claims, "The natural world is the whole of reality that we know and interact with," and no "supernatural or spiritual realm" shows up within this world, "not even in the life of humans."[18] Here, then, we can identify a seeming tension between Drees's agnosticism, and emphasis on limit-questions, and his assertion that this world and its basic constituents are purely natural. If science is limited in the way that Drees suggests, then what reasons do we have for thinking that contemporary science supports a naturalistic view of the world, as opposed to other ontologies and metaphysical systems? It seems as if Drees's willingness to recognize the intrinsic limitations to empirical enquiry clashes with his ambition to uphold a strict naturalism. Drees's epistemological assertion that science will not remove the "horizon of not knowing" seems to undercut the idea of naturalism as superior to other ways of making sense of this world. On his own epistemology, it does not seem that naturalism stands victorious.

In chapter 3, I discussed the ways in which human freedom and subjectivity pose significant problems to Drees's naturalistic project. A main conclusion was that Drees had not successfully spelled out how such phenomena can be rendered compatible with his naturalist stance and the primacy of physics. Rather than grounding freedom and subjectivity in the natural, Drees appealed to a seeming correlation between mental and physical events. In response to this, I argued that such appeals to correlation fail to ontologically explain the nature and cause of such phenomenological features. Mere correlation does not explain human mentality. However, in light of such "hard problems," Drees argues that "it is too early to give up" on a properly naturalistic theory of human phenomenology, including consciousness.[19] Science might be able to solve these riddles further down the road, Drees suggests. However, it looks as if Drees's epistemological affirmation of the limit-questions of science undermines such hope. Once again, it seems as if Drees's hope in the future verification of the naturalistic story about the world and humanity is undercut by his own recognition that science bumps up against unsolvable limit-questions. It should be emphasized here that such limit-questions are not merely unsolvable in the present, but they are unsolvable in principle. They do not merely point toward the limits of our current knowledge, but instead reveal the limits of the scientific endeavor. Drees's invitation to mystery undermines, thus, the status of naturalism. His gesturing towards an "honest agnosticism" is admirable, yet it poses a challenge to the naturalistic project, and it remains unclear on Drees's argument why we should prefer a naturalistic picture of reality over non-naturalistic metaphysical systems and theories. In this way, it seems as if the mysterious character of reality—a core feature of Drees's religious proposal—clashes with the ambition to derive a naturalistic picture of the world based on contemporary science. Given the inherent epistemological limitations to the enterprise of science, and the problems of grounding various phenomenological features of human experiencing, it remains unclear on Drees's view why we should affirm ontological naturalism. Drees's sober recognition that science is limited sets him apart from the reductionist spirit and scientistic hubris of other naturalists. Yet such recognition spells bad news for the coherence and epistemic superiority of naturalism in the landscape of alternative ontologies and metaphysical stories.

Pragmatic Realism

Several religious naturalists—such as Karl Peters, Gordon Kaufman, Stuart Kauffman, and (to some degree) Ursula Goodenough—propose a pragmatic understanding of religion, whereby religious discourse is not truth-orientated but seeks to motivate us to adopt ecologically mindful attitudes.

I applaud these religious naturalists for their emphasis on the responsibility of traditional religions, and new articulations of religious beliefs and practices, in responding to the ecological challenges. It should not be denied that certain strands of traditional religions have tended to devalue nature and have construed nature as a rival to God. What some religious naturalists show, therefore, is that religious communities need not be a part of the problem but can be a part of the solution. Thus the "great wisdom traditions," to use Goodenough's term, should not quickly reject the critique offered by religious naturalists but consider it an invitation to critically reflect on some of the predominant conceptions of the Divine or God. Indeed, I concur with Michael Hogue: "The moral dimension of religious naturalism is one of its greatest strengths in a time of ecological decline. Religious naturalist virtues and imperatives are calibrated to the scale of vision and concern needed to see and respond to the globally diffused and temporally delayed nature of ecological vulnerability."[20] However, despite some of the promises of the ecotheological vision of religious naturalism, it does highlight some critical issues in the way that it understands religious truth claims. I want to raise two concerns regarding this ecopragmatic model for theology:

1. The pragmatic construal of theology seems to entail epistemic reductionism in the area of theology.
2. It is not obvious that pragmatism implies a rejection of an ontological or extranatural conception of God.

Religious naturalists of the pragmatist inclination urge us to reconceptualize religious images so as to make us more inclined to adopt attitudes beneficial for the well-being of the ecosystem. In this way, the truth of religious images will depend on their ecological adequacy. My first concern is that this will invite epistemic reductionism in theology. In other areas, such as science, we take into account several factors when judging the validity of a certain theory. We might see if the theory in question exhibits explanatory power, whether it can help us to predict certain outcomes, if it is testable or falsifiable, and so on. The reason we need to take into account all these factors is that one set of criteria will simply not suffice in light of the multidimensional holistic nature of reality. My worry is that some ecopragmatic thinkers, given this strong commitment to ecological stability, run the risk of reducing theology to purely ecological concerns. Hence, they neglect the multidimensional nature of human spirituality. As Mikael Stenmark argues, the pragmatist accommodation of theology mixes "in an awkward manner factual matters with moral and political considerations."[21] In chapter 1, I described the ways in which religious naturalists, indeed the large majority of them, adopt realism in regard to the practice and pursuit of contemporary science. Yet it could be the case, which Stenmark elaborates on, that certain beliefs in the scientific community are politically and ecologically problematic, in that they could produce negative consequences for society and the environment. Stenmark writes, "Scientists believe, for example, that atoms can be split into parts. They also know that this belief was crucial in the construction of the atomic bomb. Some of us think that atomic bombs and other nuclear weapons are evil things that constitute a massive threat to the creation of an ecologically sustainable and peaceful society."[22] Thus, the belief that atoms can be split into parts might carry negative consequences for the future of nature and humanity. Yet most of us would still concede that it is rational to believe that atoms can be split into parts, despite the possible negative consequences of believing such things. Conversely, "the fact that it would be politically or morally desirable that p is the case does not entail that it is true or probably true that *p* is the case."[23] Religious naturalists sternly defend the epistemic authority of science and would most likely not embrace such pragmatism about the nature and practice of science. Given that religious naturalists do not adopt the same pragmatist spirit in the arena of science, it remains unclear why we need to rethink theological truth-statements on the basis of a pragmatist conception of truth and rationality. Again, the religious naturalists' ambition to reduce theological truth, but not scientific truth, to issues of ecological soundness and adequacy is strikingly problematic, and even question-begging.

Of course, some religious naturalists might respond with the argument that pragmatism and the impossibility of determining religious truth values lead to an ecotheological position, and a rejection of God as an ontologically distinct being. Given pragmatism, a strong focus on (and perhaps even reduction to) ecological concerns follows naturally. An example of this line of reasoning comes from Karl Peters, who has argued that pragmatism inevitably leads to ecotheology.

For Peters, who take his cue from Charles Sanders Peirce's pragmatic maxim, pragmatism leads to a concept of God as a nonsupernatural "immanent creative process that is coextensive with but conceptually distinguishable from the world."[24] For Peirce (and I have also examined this in chapter 4), we determine the meaning of abstract ideas through their practical effects on the lives of human beings. Peters suggests, "If we apply Peirce's way of defining meaning in terms of habits of actions and sensible results to theology, we will be led to speak of an immanent God."[25] This is because the notion of a supernatural God, a being existing beyond the natural order, cannot be meaningfully conceptualized in human discourse: "Meaningful discourse is possible only in terms of a universe open to human action and experience."[26] Peters further suggests that an immanent notion of God, which is what a pragmatic conception of meaning requires, invites ecological awareness, as God is now grounded in the natural order. God is part and parcel of physical reality, given pragmatism.

Why, however, should we take pragmatism to imply a negation of theological realism? Peters suggests that given the pragmatic maxim—that we should judge beliefs in terms of their practical effects—we cannot meaningfully conceptualize supernaturalistic beliefs. But why should we take this to support a rejection of an ontological conception of God? Is the pragmatic maxim, and the emphasis on human conceptualizations, not equally compatible with, for example, *religious agnosticism*? One could say that according to pragmatism we cannot meaningfully conceptualize beliefs in the supernatural, so we should remain *agnostic* with regard to claims to the supernatural, while retaining the possibility of engaging reality religiously.

This line of reasoning would come fairly close to John Hick's Kantian position. Perception, argues Hick, is always interpreted. The fact that "the mind imposes order and meaning upon the data it receives is true at all levels of awareness, physical, moral, aesthetic and religious."[27] In what I suggest is a pragmatist fashion, Hick explicitly recognizes the contextuality and situatedness of human creatures. Human experience is always "experiencing-as" (Hick develops Wittgenstein's notion of seeing-as), whereby the mind is actively interpreting its environment and sensory information by "means of concepts and patterns drawn from its memory."[28]

Thus Hick suggests that in the realm of religion, there will always be a difference between the "Real in-itself" and the "Real as experienced by religious believers." Our concepts refer to "an ultimate divine Reality which transcends all our varied visions of it."[29] Hick, in his Kantian model, concludes, "We now have to distinguish between the Real *an sich* and the Real as variously experienced-and-thought by different human communities."[30] This Kantian model, as was seen in the previous chapter, was also central to the earlier work of Gordon Kaufman.

Comparing the pragmatist model of religion proposed by some religious naturalists with the Kantian/critical realist model of religion, we can observe the following: while both the pragmatist and the Kantian realist stress the contextuality of religious believers and the problem of referring to reality in itself, they draw radically different conclusions from this regarding the ontology of God or the ultimate/Divine. The pragmatist denies the notion of an ontologically distinct being, while Hick's pragmatic realism retains the realist element of religion but adopts an agnostic attitude toward the ultimate reality. My critical question is therefore, why should we prefer the religious naturalist denial of an independent (supernatural) religious reality to the religious agnostic position?[31]

I further want to suggest that pragmatism need not imply a rejection of religious realism, or more specifically the denial of a transcendent God. It all comes down to our definition of "pragmatism." Here I want to turn to Nicholas Rescher, who has sought to restore the realist dimension of Peirce's pragmatism. He asks us to consider two very different versions of epistemic pragmatism. The first is a deflationary and deconstructionist version of pragmatism, which urges us to forget about truth, as the pursuit of truth is pointless and illusory. The second instead retains the notion of truth (as correspondence to reality) and seeks "to guide us on its rational pursuit" by offering certain criteria for theory testing.[32] Rescher argues that pragmatism is a way of understanding the process of justifying theories. Hence, pragmatism as a method does not negate, for example, metaphysical realism or ontological conceptions of truth. I suggest based on this Rescher-inspired approach that pragmatism, when properly construed, does not entail a deflationary account of truth and is therefore not a negation of religious realism. Indeed, Rescher sees no conflict between his objectivist pragmatism and his own process-theological position.[33] We can hold to both a pragmatic understanding of religious praxis and discourse while retaining the realist intention of religion, and a realist understanding of the truth claims made in the religious sphere. Ecological concerns can perhaps enter into the justification phase of religious images and symbols, but they should nevertheless not be considered the only relevant epistemological factor. We do not have to deny theism in order to develop ecologically mindful attitudes, and pragmatism allows for a theistic approach that takes the commitment to ecological stability seriously.

Functional Religion and Theological Realism

I have argued that for Hardwick there is a major tension between his physicalism and Christian discourse. Drees's monism, however, creates a real possibility for responding to reality in a religious way. Yet Drees's emphasis on limit-questions and the intrinsic limitations of science seem to undermine the plausibility of a purely naturalistic understanding of the world. I have also called into question the pragmatist rejection of theological realism. We shall now turn to Loyal Rue, who thinks that realism in religion is not only possible but a necessary presupposition. Is Rue's realist religious naturalism philosophically adequate, and is religious realism compatible with his overall evolutionary naturalism?

As mentioned earlier, Loyal Rue argues that religion, in order to function properly, needs to be understood realistically. Consequently, anti-realism undermines the function of religion, which is to produce spiritual transformation. I agree with Rue to a large extent on this issue, and support his defense of religious realism. However, can Rue, with his own model, uphold realism regarding religious discourse in general and the "religious root metaphor" in particular"? Although I applaud his desire to uphold religious realism, I doubt the successfulness of this project.

Religion, as understood by Rue, is not about God. Instead it is about personal wholeness and social coherence, and these two are dependent on the integrity of the biosphere as a whole. Speaking in evolutionary terms, Rue argues that religion is about the maximization of *viability*, the odds of surviving. In this way, the "ultimate function of a religious tradition is to enhance personal wholeness and social coherence by nurturing the conscious and unconscious lives of individuals."[34] Hence, whenever a functional breakdown in religion occurs we "observe systematic losses in the achievement of personal wholeness and social coherence."[35] Religion, then, is not about God or some supernatural aspect of reality but about maximizing our viability and so ensuring the survival of the human species.

Rue proposes an interesting evolutionary account of the function of religion. However, I suggest that Rue, by virtue of reducing religion to biological viability, actually undermines the realist intention of his religious naturalism. According to Rue's own model, we should conclude that religious naturalism is not about the sacredness of nature, because its main function is simply to increase the odds of surviving. If we take the logic of Rue's project seriously, the perceived sacredness of nature is not about nature itself; it is an adaptive myth. That is, in the same way that religion is not about God, religious naturalism is not about the sacredness of nature. Rue is replacing realism with an evolutionary reduction so that the function of religion is not to produce spiritual transformation but to secure the survival of the individual and the collective.

In one of his earlier works, Rue indeed suggests that the meaningfulness of life is an illusion, and that deception plays an important part in both personal wholeness and social coherence. Deception "occurs when a discrepancy between appearance and reality can be attributed in part to the causal influence of another organism."[36] This phenomenon, argues Rue, is not confined to human beings. On the contrary, deception is widespread throughout the natural domain, at various levels in the organization of life, including molecules, plants, insects, aquatic animals, mammals, and so on.[37] Rue further argues that deception has been, and still is, a vital function in the process of evolution. Rue writes, "When deceptive traits appear in nature, they tend to be favored. These traits then become significant elements in the process of natural selection; that is, the presence of deceivers bears upon the fitness of their dupes."[38] However, in all of nature, human beings are the most resourceful and skilled deceivers.

As I have mentioned earlier, Rue takes personal wholeness and social coherence to be the two primary aims of human beings. It is suggested that certain deceptive strategies are relevant for the attainment of personal wholeness. It is often held that self-deception is a paradoxical phenomenon, but Rue claims that such an occurrence is natural when "one considers a person to be a set of distinct yet interacting motivational processes."[39] An example of deception includes, according to Rue, the construction of a strategy to leave the deceived, or the duped, in a state of ignorance so "that the result is enhanced social esteem for the deceiver."[40] Strategies of *self*-deception mean that the self-deceiver formulates and employs a particular strategy to prevent certain information from reaching consciousness.[41] Strategies pertaining to the attainment of personal wholeness also involve the intention to deconstruct and evade threats of self-annihilation by blocking out information so that a potential threat to one's sense of safety is diminished.[42]

Deception is also crucial for the stability and coherency of society, and for social groups that share an identity, whether it be a political party, a religious group, or a sports team.[43] A group that lacks coherence would lack a system of widely shared meanings and beliefs that "serve to integrate and regulate the culture as a whole."[44] Rue explains that social coherence is dependent on a delicate balance between deception and honesty; too little or too much of either means that a breakdown in social stability of a particular group or culture will occur. Deception is common in "primary groups," such as families and marriages, in "secondary groups," such as business groups, and in "comprehensive groups," for instance nation-states (who seek to maintain their coherence through war and diplomacy).[45] In all these cases, the group has to find its particular pattern for managing the balance between deception and honesty, and deceptive strategies "play a constructive role in establishing confluences of interests."[46] It is with regard to social groups that Rue emphasizes the binding power of myth and, thus, religious traditions. Religious myths can increase social coherence by providing a system of shared meaning and beliefs, a narrative of cosmology, economy, and ethical guidelines and morality.[47] Rue suggests that deception may "advance the efforts of religious authorities to establish and maintain social coherence in a comprehensive group."[48] Deceptive strategies within the domain of religion also include the construction of particular conditions for "adherence to the myth," as well as the deconstruction of conditions "for emergent rival authorities."[49]

Interestingly, Rue concedes that nihilism is true. To say that nihilism is philosophically true "is to say that all meanings are contingent caricatures of reality, and that no ultimate account of the way things really are can be universally enforced."[50] Nihilism is true, but evolutionarily speaking it is a *maladaptive* truth, according to Rue. According to this nihilism, all myths are false, but "without myths human beings lack the resources to achieve personal wholeness and social coherence."[51]

Rue goes on to argue that if we want to avoid the nihilist conclusion that the universe is dead and void of meaning (which would threaten personal wholeness and social coherence), our only option is to construct a "noble lie." This noble lie must be more than plausible—it must be adaptive. A noble lie, even though objectively false, can provide a level of moral guidance and spiritual fulfillment, and ensure social cooperation. What humanity needs, according to Rue, is a *new naturalism*, which is the general outlook that has emerged in light of the advancements of science. This naturalism is vulnerable to the nihilist threat, but it can still provide a *story* that, centered on the evolutionary origin of all of life, can unite the dispersed cultures of humanity. The new naturalism, however, "may not be everybody's story in the absolute and final sense, but it is far and away the closest anybody has come. It is more noble."[52] The survival of humanity depends on a shared myth, and in particular a *biocentric myth*, an ecologically informed myth that focuses on the "conditions for a rich diversity of life forms."[53] This myth also affirms the integrity of nature, and that nature is the ultimate source of life. A biocentric myth enables human flourishing, and the achievement of personal wholeness and social coherence. Rue is clear on the fact that it is a lie, but it is a *noble* lie, as it helps people to more fully enjoy life, despite the harsh truth of nihilism. Rue writes, "Biocentrism is your story and mine. It is everybody's story. It presumes to tell us how things are and which things matter. It is, nevertheless, a lie."[54]

Rue offers an interesting way to "bring religion down to nature." Yet given his reductionist explanation of religion in terms of reproductive fitness, his view of viability as the only moral standard, and his view of a biocentric spirituality as a lie, a radical revision of religious discourse is entailed. This radical revision implies some worrisome implications for Rue's desire to uphold religious realism.

In departing from a realist understanding of religion and effectively adopting a view of religion as a noble lie, Rue's proposal comes fairly close to what is generally referred to as "religious fictionalism," whereby religious discourse is understood as useful fiction.

I suggest that Rue's evolutionary understanding of religion is vulnerable to the critique made by John Hick against naturalism: that when the bleak implications of naturalism are recognized—that meaning in life is a matter of pretending—naturalism turns out to be "grim news for the many."[55]

This dimension of Rue's naturalism should be recognized in order to fully appreciate the problem of this fictionalist stance. The British biologist Alfred Russell Wallace notes the existential problems produced by a materialist account of evolution when he writes,

> We, in common with the rest of nature, are but products of blind eternal forces of the universe, and believing also that the time must come when the sun will lose his heat and all of life on earth necessarily will cease . . . we are compelled to suppose that all the slow growths of our race struggling towards a higher life, all the agony of martyrs, all the groans of victims, all the evil and misery, and undeserved suffering of the ages, all the struggles of freedom, all the efforts towards justice, all the aspirations for virtue and the wellbeing of humanity, shall absolutely vanish.[56]

Wallace acknowledges, much like Hick, that a materialist account of evolution entails the meaninglessness of human striving and the pointlessness of physical existence. Human life, of course, contains much joy as well, but, as Hick notes, "too much of this realm of good experiences is available only to those that have been lucky in the lottery of life."[57] Indeed, even those people that have lived the longest "can seldom be said to have arrived, before they die, at a fulfillment of the human potential."[58] Rue himself suggests that we will not respond to the ecological crisis in time and that a doomsday scenario is inevitable.[59] I am not making the inference that because Rue's evolutionary reductionism invites nihilism (and ends up denying the purposefulness of human life), it must be false. Instead, I am suggesting that Rue's fictionalism becomes strained when one acknowledges the existential conclusions that follow from this evolutionary account of meaning and purpose.

In the end, Rue's construal of religion, whereby the function of religious beliefs is to aid in achieving reproductive fitness, seems to rule out religious realism. He seems aware of this when he writes, "And religious naturalists may affirm the sacredness of Nature and practice eco-centric piety sincerely, yet deep down they must know that religion is no more about Nature than it is about God."[60] Rue has argued that the function of a mythic tradition is dependent on a realist self-understanding. However, Rue's own biocentric myth seems to invite the same "creeping nonrealism" that he fears will undermine the important function of religious traditions.

Objectivist Religion and the Problem of Evil

Now we shall consider an alternative realist and objectivist conception of religion, offered by Donald Crosby. For Crosby, the religious striving within naturalism is not a matter of pretending, or something that we come to engage in because of some potential pragmatic benefits. Nature, as the religious object, fulfills the same role as "God" in traditional religion (by fulfilling the role-functional criteria outlined in chapter 4).

Moreover, Crosby maintains that religious devotion should not be restricted to a certain aspect of nature. Rather it should be directed toward nature in its entirety. I will argue that Crosby faces the challenge of the ambiguity of reality, namely that nature contains both good and evil, both flourishing and suffering. The problem of evil/suffering stands as a severe challenge for several forms of religious naturalism, and it would certainly play out differently depending on whether one embraces the whole of nature as religiously significant or one maintains that some specific aspect of the natural order is worthy of religious affection.

As Crosby has given extra attention to this issue, and provided a sophisticated defense of religious naturalism in light of evil and suffering, I will discuss his proposal at greater length. Indeed, apart from the metaphysical problems facing Crosby's emergentist metaphysics (see chapter 3 regarding the problems of pluralistic naturalism), evil/suffering constitutes the greatest challenge for his "religion of nature."

Crosby is acutely aware of the fact that there *is* a problem of evil for religious naturalism, or "religion of nature," as he prefers to call it. Nature contains awesome beauty, but it also contains destructiveness: "tranquility and suffering is there."[61] How is it possible to revere, to devote one's trust, to a religious object that contains much pain and suffering? Crosby maintains that the ambiguity of nature, the beauty, cruelty, and destructiveness of nature, is not a problem for his religious proposal. On the contrary, the reality and unavoidability of ambiguity speaks in favor of an immanent religiosity: "Ambiguity is built into any robust and genuinely desirable world, then, and our natural world—despite and even *because* of its ambiguities—is worthy of our utmost religious trust, devotion and commitment."[62]

It seems as if Crosby bases his religion of nature and the plausibility of it on three key assertions:

1. Reality is necessarily ambiguous, and a religious object does not have to be unquestionably good.

2. We should separate moral goodness from religious rightness.
3. Nature can provide existential assurance.

These three argumentative claims will be outlined and responded to below.

Necessary Ambiguity and Perspectival Evil

Crosby argues that there is no such thing as a (morally) perfect world. There are both *systemic natural evils*, such as pandemics, earthquakes, and floods, and *moral evils*, such as genocides and mass murders. To imagine a perfect world is to imagine a world in which there is no scarcity of resources, no hunger or thirst, and where no sentient being would prey on other creatures. Moreover, "All its human beings without exception would do only good and never evil."[63] The problematic consequence of such a world, according to Crosby, is that by negating all evils in nature, we are also negating the goods. Nature's creativity would be absent in a perfect world, as nature's "creations are bought at the price of its destructions."[64] The creative aspect of nature is always in transformation; hence, when something new comes into being, something else has to be left behind. This is why the good of biological diversity is dependent on the struggles between species, leading to the death and extinction of some. Crosby further argues that if there were such a thing as a perfect world, there would be no place for human freedom, as it would have to be causally determined such that no evil action could be carried out. Humans would be like smoothly running machines, preprogrammed to do good deeds. However, this would no longer allow for free inquiry, and we would have no way of knowing what is true. Indeed, genuine freedom would be lost in this world. Crosby therefore concludes that a morally ambiguous world contains more goods than an imagined perfect world, and that "reality and ambiguity go necessarily together."[65]

Crosby claims that just as there is no nature in itself, there is also no intrinsic evil. Something becomes evil when it is *experienced* as evil by a sentient being. Thus someone who suffers from starvation, diseases, natural disasters, and so forth justifiably sees such things as evil when they are experienced as such. Crosby's idea that there is no such thing as a thing-in-itself, and that something acquires its "itness" through the perspective of sentient beings, is called "perspectivism." This move away from pure objectivism, argues Crosby, is not to succumb to relativism or subjectivism. Rather, it is the affirmation of relationalism, meaning that the world we encounter is a world of relations, and each person's perspective is bound to be different. He writes, "The world and each and every aspect of it are neither completely external nor internal. They are relational."[66] Consequently, both evil and goodness become relational terms and can never be judged independently of someone's perspective.

Crosby has offered an interesting and sophisticated account of evil, its relationship to a religion of nature, and how nature can still be considered religiously good despite the prevalence of both moral and natural evils. One of the major assumptions that Crosby takes issue with, and that he needs to undermine in order to establish a religion of nature, is the idea that a religious object needs to be morally unambiguous. Crosby's first step in this argument is to show how classical theism and modified theism (like process theism) cannot account for evil. According to these theisms, evil is a metaphysical anomaly. For a religion of nature, however, instances of evil are a necessary consequence of the life-giving forces of nature. That is, goodness will always be accompanied by suffering and evil, as the balance of the ecosystem is dependent on both creative forces and destructive forces. Hence, there is no perfect world, and a religious object must necessarily involve moral ambiguity. I have some reservations regarding Crosby's line of reasoning.

First, the supposed failure of theism to account for evil, and the unavoidability of evil and suffering in a religion of nature, does not lead to the conclusion that a religious object per se must be morally ambiguous; the only thing this argument shows, so far, is what is already stated, namely that theism has a problem of explaining evil, and evil/suffering is a necessary feature of a naturalistic religion. Crosby needs to provide some additional reasons if we are to join him in thinking that an object worthy of religious faith does not have to be unquestionably good. Otherwise, one could still maintain that a religious object needs to possess full moral goodness, and simply conclude that neither traditional theism nor religious naturalism can properly account for evil.

Second, Crosby's treatment of theistic theodicies and the free-will defense, as well as the process-theistic response to evil, is unfortunately rather superficial. The free-will defense, according to Crosby, suggests that the reality of evil is "worth the price if the goods of human freedom and personhood are preserved."[67] But, he goes on, "God's presumed unqualified, indisputable moral goodness is compromised or at least made seriously questionable by the stark evils he permits."[68] Now, the thrust of the free-will defense, at least as formulated by Alvin Plantinga, is to show that (a) a world containing free creatures is more valuable than a world containing no free creatures, and (b) that God can create free creatures but cannot determine such creatures to do only what is right and good.[69] In this way, God's omnipotence, omniscience, and goodness are preserved, regardless of the fact that reality contains a vast amount of suffering and evil. Consequently, if Crosby is going to undermine the free-will defense, and so introduce ambiguity within this theistic picture, he needs to either undermine and say that (a) the value of free creatures does not outweigh the evil in reality, or (b) that God can accomplish something that is logically impossible. It is not enough, then, to simply say that the evil in reality somehow undermines the moral goodness of God. Given that Crosby seems to affirm (a), it is up to him to show (b), that God should be able to achieve the logically impossible.

Crosby departs from a more traditional realist conception of evil by claiming instead that certain acts or happenings become evil if they are experienced as evil by sentient creatures. This goes rather nicely with Crosby's overall naturalistic metaphysics, as a metaphysically realistic account of evil might invite a form of platonic dualism or supernaturalism. When I say that Crosby does not adopt realism with regard to ethics and morality, I am not claiming that his proposal *necessarily* introduces antirealism or relativism, although I think that his perspectivism, as presently outlined, does not stand too far apart from such approaches. The benefit of perspectivism is that it might be able to avoid overly metaphysical approaches to ethics. However, it seems that perspectivism is too weak, at least with regard to the pragmatic goal of Crosby's religion of nature, which is to uphold and secure the value of the ecosystem and its inhabitant creatures.

Crosby defines evil in terms of the experience of pain, such that there is a necessary relationship between the two. That is, wherever we have pain, there is evil, thus pain is evil. Crosby needs to show that pain is a necessary and sufficient condition for evil, and, perhaps, that there is a correlation between the amount of pain and the amount of evil. Most of us, however, would immediately see upon reflection that pain, or suffering, is not always evil. Let us say that I take my cat, Kiwi, to the veterinarian because of an ear infection. At the veterinarian's office, they clean her ear and put in some bacterium-killing eardrops. From my perspective, I am doing this to help my cat, and thus I am doing something morally good. However, my cat, who does not realize that the ear drops will make her feel better, and who feels distressed by the ordeal, will fail to recognize the benefits of this visit. Perhaps my perspective is the right one, and Kiwi misjudges the situation.

However, Crosby writes, "They [the events of nature] are intrinsically bad from the perspectives from these beings, no matter how they may be seen from other perspectives."[70] Consequently, if we follow Crosby, we have to say that my perspective and my interpretation of the situation is not any truer than Kiwi's. Hence, it is very hard to see how Crosby's metaphysical and perspectival pluralism does not entail some sort of relativism.

Crosby could respond by saying that there might be some way of judging the truth value of competing perspectives. But, if that is the case, then Crosby would go beyond his original perspectival account of evil, given that additional factors other than the experience of pain and suffering would now be relevant for establishing the ontology of evil.

Another question invited by Crosby's perspectivism concerns the adequacy of his ecological ethics. Crosby's aim is to show that sentience and mind are widespread in nature, and not confined to human beings. Therefore, we are less likely to treat nonhuman animals and insects as "its" and more like "thous," and we can in this way enter into an I-thou relationship with nature. Drawing on Evan Thompson, Crosby argues that many forms of life show clear signs of sentience, intentionality, and purpose.[71] This is what defines any living system; hence, where there is mind, there is life, and vice versa (and where there is no life, there is no mind). However, Crosby also points out that mind is not ubiquitous in a strong sense but is restricted to those living organisms with a sufficient level of organizational complexity.[72] The inwardness of all forms of life should make us regard and treat nonhuman animals and organisms "with reverence as sacred beings of incalculable worth and significance."[73] All forms of life have intrinsic value and should be treated with the utmost dignity.

Even though I admire Crosby's thou-focused and eco-ethical approach, I see potential problems. It is worth remembering that Crosby, unlike other proponents of religious naturalism, regards the whole of nature as religiously significant. Crosby argues that we can enter into an I-thou relationship with nature, yet he confines mind (or "thouness") to certain creatures and organisms. This would seem to negatively affect the power of Crosby's eco-ethical vision, as we would be unable to enter into an I-thou relationship with many aspects of nature. Hence, not all of nature would be included in this eco-ethical vision. If the intrinsic value of a phenomenon is dependent on that phenomenon possessing sentience, then we must wonder what happens to those organisms and aspects of nature that cannot be considered as "thous." Crosby seems to have two options: he can either adopt instrumentalism with regard to the nonsentient aspects of nature or turn to panpsychism/panexperientialism by saying that mind reaches all the way down in nature.

The instrumentalist approach is indeed a coherent approach but seems incompatible with Crosby's overall religious vision. On this view, the non-sentient aspects of nature would have value only insofar as they contribute to the well-being and flourishing of sentient creatures. However, if some aspects of nature now have only instrumental value, then that would seem to undermine the religious significance of nature as a whole, given that certain features of physical reality are clearly more valuable than others.

What about the panexperientialist or panpsychist direction? What if Crosby extended the thouness of nature to include all of nature? Crosby (as seen in chapter 2) views reality in terms of emergence theory, and has also critiqued panexperientialist views proposed by Frederick Ferré and Donald Dombrowski.[74] I will argue in the final chapter that a panpsychist/panexperientialist direction is a fruitful one for not only Crosby but other religious naturalists as well.

Moral Goodness and Religious Rightness

Having looked at the first of Crosby's three arguments, we now move on to the second. Crosby defends a religion of nature from the problem of evil by challenging an often-held assumption, namely that an object, to be an appropriate focus of religious faith and dedication, must be unqualifiedly good in the moral sense of the term. This assumption is strongly critiqued by Crosby. What Crosby is arguing is that although nature is *morally* ambiguous, it is still *religiously right* to worship nature and to adopt an attitude of awe and wonder toward it. This is because religion and morality, in many ways, are significantly different.

Morality, argues Crosby, is dedicated to teaching us the difference between goodness and evil through appropriate instruction and cultivation. Therefore, morality is concerned with things that are under our control. This is not the case for religion, as its main focus is on "the events of grace," the unexpected events that lie beyond prediction and control, and the possibility of transformation that such events provide.

Another difference, Crosby continues, is that morality, unlike religion, is concerned with only a limited part of reality. Religion relates to the whole of life and the universe, and it provides a view of what is ultimately significant in life. Moreover, religion provides a framework for understanding how human beings fit into the scheme of things. Religion, given its broader scope, can show how morality relates to different areas of life. Crosby concludes, "Moral values are not equivalent to religious values, and moral problems and concerns are not the same thing as religious problems and concerns."[75]

The idea so far is that we should be careful to differentiate religion from morality, that moral concerns are not to be equated with religious concerns. It is also argued that religious rightness and moral goodness are different. What does the term "religious rightness" amount to? Crosby suggests that it means "living gratefully and responsibly" and "giving ourselves for the well-being of the earth and all its constituents and systems."[76] Hence, when we give thanks to nature for the gifts of consciousness, intelligence, and freedom, we are *practicing the rightness of nature*. This practice also means that we should work in our respective religious communities for the betterment of the world, while accepting the inevitability and even necessity of natural evils. Crosby therefore concludes, "Understood in these ways, the ambiguities of nature present no barrier to its being regarded as the focus of religious faith."[77]

What does the distinction between religious rightness and morality actually accomplish? I am honestly not sure what Crosby seeks to achieve with this distinction, but he seems to be saying that if we can differentiate religion from the moral realm, then we can do away with the tension between a religious outlook on life and the experience of natural and moral evils by sentient creatures. But why should one think that? Just because religion and morality are not identical does not imply that moral matters do not have any impact on religion or the existential adequacy of particular religious objects. To establish the difference between religion and morality does not answer the main question, which is, "Should I feel gratitude toward a religious object that also contains much suffering and evil?" The classical theist, or the supernaturalist, can easily concede this difference between morality and religion. Nevertheless, the theist is still forced to engage the philosophical and existential problem of worshipping a God who seems to allow evil and suffering. The religious naturalists, likewise, are forced to engage the issue of revering an object that is simultaneously the ground of goodness and evil.

Crosby could, however, maintain a stronger thesis regarding the relationship between morality and religion, and so escape the problem of having to explain, or perhaps justify, the reality of evil from a religious standpoint. He could completely separate the religious domain from the moral domain. In this way, a religion of nature would remain insoluble to the moral problems, as the function of religion would be categorically different from that of morality. Therefore, we could without inconsistency conceive reality religiously regardless of suffering and evil. However, this is not an attractive option for the Crosby-type religious naturalist. According to Crosby, who outlines several ethical demands that he thinks flow from a religion of nature, moral obligations "of various kinds are critically important in this religious vision of rightness."[78] He states also that we must "work within the whole of nature to bring about moral good and to avoid needless suffering and harm."[79] Morality, therefore, is an intrinsic part of a religion of nature, not an optional add-on, and so to separate religion from morality completely is, in fact, not an option for Crosby.

Coping with and Mitigating Evil: Assurance and Demand

Crosby's third argument, as outlined above, states that despite moral and natural evils, nature provides assurance. Because of the assurance that nature offers to us, we have a way of coping with the destructive dimension of reality. This is connected to Crosby's naturalistic soteriology. However, what does Crosby mean by "assurance"? For many traditional believers, the religious belief in postphysical survival provides a particular kind of existential assurance. This is not the kind of assurance that Crosby has in mind, given what he believes are insurmountable problems for supernaturalism. What Crosby is arguing is that the assurance aspect of a naturalistic soteriology flows from the "conviction of intimate connection with nature and of sharing with all its other creatures a settled sense of belonging . . . and of being completely at home—albeit in a precarious world."[80] We are thus "healing the aching sense of homelessness and alienation created when one views oneself as being something other than a natural being."[81]

All the potentials bestowed on us by nature can be steered in both evil and good directions. With these possibilities, argues Crosby, come responsibilities and demands. We have a responsibility to the well-being of nature and other creatures; we must make sure to work against the mistreatment of animals, avoid needless despoliation of the earth, and make an effort to conserve nonrenewable sources of energy and to "treat all living beings with respect and to reverence the earth as our natural home."[82]

However, as I argued with regard to Rue, when the full existential consequence of naturalism is recognized, the optimistic view of nature can be called into question. Nature is not, as many naturalists recognize, a purely harmonious system. There is much suffering and evil within the natural order. Crosby recognizes this aspect and sees it as crucial for his own proposal. Crosby further concedes that his naturalistic view does not secure the triumph of good over evil, "and there certainly can be no possibility of the elimination of evil altogether, given the ambiguous relations of good and evil."[83] What happens, however, to the meaningfulness of life for those groups and individuals who do not merely suffer but suffer due to irredeemable evils that do not in any way bring about some corresponding good?

It should be pointed out that Crosby's perspectivism prohibits him from claiming that despite the experience of horrendous evils, life still works out for the greater good. Crosby's perspectivism means that *experience* is the ultimate criterion for judging whether something is evil, hence this naturalism cannot provide any independent assurance in light of the experience of evil. This does not mean that Crosby's naturalism is false, but it raises serious problems concerning the assurance aspect of a religion of nature.

The Many Religious Objects of Religion of Nature

Above I described and critiqued Crosby's three key assertions: that a religious object must be morally ambiguous, that the distinction between morality and religious rightness offers a way forward for a religion of nature, and that nature provides existential assurance. In the following and final section of this chapter, I question whether Crosby consistently holds to the idea that *all* of nature should be considered religiously significant and worthy of reverence and awe.

Crosby contrasts his own view with that of Henry Nelson Wieman. For Wieman, God is seen as the "Source of Human Good," and more specifically as a radically immanent process within nature.[84] God is the creative good that gives rise to qualitative meaning. This is also called the "creative event," which is constituted by four subevents: the creative event producing new value due to the enriching communication between organisms; the internal integration of new meanings with meanings previously acquired; the organism's appreciation of the world necessarily being expanded and enriched if new meanings are successfully integrated; and those who participate in the total creative event experiencing a widening and deepening of the community.[85]

Like Crosby, Bernard Loomer views God as the totality of the world. Loomer critiques Wieman's conception of God for lacking adequate size: "Such a God would be hopelessly too small, limited, and abstract, in contrast to the vastness, complexity, and concreteness of the dynamic world of experience."[86] Crosby concurs with Loomer's assessment.

However, even though Crosby claims to view all of nature as religiously significant, similarly to Wieman, he focuses on the goodness of the religious object. Nature, says Crosby, is the source, sustainer, and restorer of life: "The relentless powers of re-creation can inspire profound feelings of gratitude and awe."[87] Nature is also the source of the good of human life, and human aspiration toward goodness is a gift of nature.[88] Crosby, of course, does also acknowledge suffering in nature. We should for that reason expect Crosby to maintain the equal religious status of the creative and destructive forces of nature, and so uphold a more all-inclusive conception of God by saying that *all* of nature is to be revered. Yet he does not. Crosby recognizes the reality of natural and moral evils, and the suffering experienced by sentient creatures. He writes, "*Much* in nature is to be revered, and it is a fit object of religious devotion."[89] That is, not *all* of nature is to be revered, but rather those aspects of nature (or *natura naturans*) that are beneficial for human beings and the ecosystem as a whole. Moreover, Crosby writes that "the moral ambiguities of nature are mirrored in the moral ambiguities of human nature."[90] Crosby also urges us to maximize the goodness in the world and dedicate ourselves "to moral excellence and the service of others, including nonhuman others."[91] Consequently, he urges us not to align ourselves with the destructive forces of nature but with those forces aimed at increasing the goodness of our natural home. In the end, Crosby, not too different from Wieman, seems to focus the religious attention on the life-enabling and sustaining creative forces of nature. It seems, then, that Crosby, given this one-sided focus, undermines his own claim that *all* of nature is to be regarded as religiously significant.

Conclusions

In this chapter I have discussed several issues connected to the religious dimension of this emerging naturalistic perspective. With regard to the nonrealist interpretation of religion, I argued that Hardwick's religious proposal is problematic, as the Christian part of this project is grounded not in physicalism but in the Christian tradition itself. Drees's form of naturalism is more promising, as he recognizes the limitations of naturalism and empirical inquiry and so allows for a genuine communication between naturalism and religious traditions. Yet it was seen that Drees's recognition and affirmation of unsolvable limit-questions undermines the plausibility of naturalism as a sufficient story about this world.

Thereafter, I critically evaluated the pragmatic construal of religion, that is, pragmatic religious realism, proposed by Kaufman, Kauffman, Peters, and Goodenough. I critiqued this perspective for its epistemic reductionism. I also suggested, contrary to the opinions of some proponents of religious naturalism, that pragmatism does not necessarily entail a rejection of an ontological or supernatural view of God. A pragmatic construal of religion is indeed still compatible with religious agnosticism and the Kantian position espoused by John Hick.

Lastly, I evaluated two forms of religious realism, advocated by Loyal Rue and Donald Crosby. I argued that Rue's in many ways admirable defense of religious realism fails, as his evolutionary reduction of religion (in terms of reproductive fitness) seems to contradict a realistic understanding of religious discourse.

The problem of suffering/evil was addressed with regard to Crosby's religion of nature. Crosby attempts to offer an account of evil from the viewpoint of religious naturalism by affirming ambiguity as an unavoidable part in a robustly religious outlook. This, however, was deemed unsuccessful.

I have highlighted significant problems facing religious naturalism, with regard to both naturalism (chapters 2 and 3) and the religious dimension (chapters 4 and 5). I will now explore three alternative frameworks that may help religious naturalism to move forward. In chapter 6, I will explore different naturalistic ontologies in order to combat the problems outlined in chapter 3. In chapters 7 and 8, I will explore possible theistic ontologies and panpsychism, respectively.

6

Alternative Ontology 1

Naturalistic Options

Religious naturalists seek to carve out a sustainable middle path between supernaturalistic religion and full-blown secularism. The naturalistic component of this emerging perspective, however, encounters severe problems in both its monistic and pluralistic forms. Monistic naturalism collapses into crude reductionism, and pluralistic naturalism—given its more expansive ontology—entails a dualist picture of human mentality.

As a result, one might conclude that religious naturalism has failed. However, this conclusion is premature. Many naturalists would agree with my assessment that there is a metaphysical grounding problem for standard metaphysical naturalism, but instead of giving up on the naturalistic project as such, they attempt to counteract such problems and challenges by proposing new formulations of naturalism. These new formulations take us beyond both monistic and pluralistic naturalism. In this chapter we will consider three alternative understandings of a naturalistic ontology: liberal naturalism, agnostic naturalism, and pragmatic naturalism. The aim is to reconstrue the agenda of religious naturalism in light of these three positions and evaluate their potential for moving contemporary religious naturalism forward in the science-religion debate.

Liberal Naturalism

Liberal naturalism is, roughly speaking, an attempt to find space between scientific naturalism and supernaturalism. Mario De Caro, Alberto Voltolini, and David Macarthur have argued that reality is inhabited by several entities that do not correspond to the kind of entities that science deals with. These are normative/moral facts, mental entities, and abstract objects, such as numbers. The various metaphysical grounding problems facing naturalism have led supernaturalists to conclude that we must place these nonphysical phenomena in a realm beyond the purely natural. Naturalism, it has been said, has no place for these properties, and excluding such properties should make one doubt the explanatory adequacy and coherency of this framework. Liberal naturalists hold to the constitutive claim of naturalism, namely that we should never accept an entity or explanation that *goes against* the laws of nature.[1] Still, unlike other (reductive) naturalists, one who embraces a more liberal view also claims that "there may be philosophically legitimate entities that, on the one hand, are *ineliminable* and, on the other hand, are not only *irreducible* to scientifically accountable entities but may also be ontologically *independent* from them."[2] The challenge, of course, is being able to uphold the theses of noneliminability, irreducibility, and ontological independence without turning this position into crypto-Platonism, dualism, or supernaturalism. A liberal naturalist, as compared with a scientific naturalist, seeks to widen the realm of the natural while maintaining the integrity of the basic causal nexus. The liberal also rejects the Quineian continuity thesis. That is, there does not have to be continuity between the methodology and ontology of science, and that of philosophy. Philosophy constitutes its own domain and can produce knowledge independently of the natural sciences. Liberal naturalism in this way becomes a real and genuine alternative to those forms of naturalism that have given in to a reductionist understanding of the world. De Caro and Voltolini write, "Ontological tolerance plus methodological discontinuity explains how Liberal Naturalism may escape the first horn of the dilemma. In allowing for entities that cannot be studied by the methods of natural sciences, Liberal Naturalism shows how it differs from Scientific Naturalism."[3] There is, of course, a minimal set of beliefs or commitments that must be shared by any robust naturalist. Macarthur pinpoints three necessary commitments: one must reject the supernatural; one must concede that human beings can be studied by the natural sciences; and one must have respect for the conclusions of the natural sciences.[4]

The liberal naturalist in this way holds to basic naturalism and affirms simultaneously, against the reductionist credo, that there are entities that go beyond the categories of the natural sciences. What kind of entities do De Caro, Voltolini, and Macarthur have in mind? De Caro and Voltolini argue that modal properties are a good example "to which liberal naturalists can appeal in order to differentiate themselves from both Scientific Naturalism and Supernaturalism."[5] Modal thought is essential to human thinking: we make judgments and we evaluate statements of *possibility* and *necessity*, and this kind of thinking is essential for theoretical and practical reasoning. Modal properties do not seem to be the kind of properties that are instantiated or individuated by any physical theory. They seem to "exist over and above scientifically explainable entities."[6] Yet modal properties can be considered real within naturalism, as they do not interfere in, or interrupt, the causal order. They are, in some sense, *natural* while resisting explication in terms of the physical categories familiar to science. Liberal naturalists, as we can see, adopt a more expansive view of what it means for higher-level phenomena to be natural; they also employ a less restrictive criteria for judging whether some property or object can be accommodated within a framework that prioritizes the reality of the natural.

Macarthur argues for a similar liberalization of naturalism, but with regard to normative and methodological truths. He suggests that naturalism so far has failed to account for normative properties; hence, many naturalists have not been able to bridge the manifest image and the scientific image of reality. Some naturalists have claimed that science must be "value-neutral," that we need to discard normativity. Macarthur says that this view is deeply mistaken, and he accuses scientific naturalists for failing to recognize the incoherency of rejecting normative properties, given that scientific naturalism *itself* is a normative doctrine; it tells us that we *ought* to admit the existence of only those things that can be explained by science, and that we rationally *ought* to acknowledge that the natural sciences are the only knowledge-producing disciplines.

Some scientific naturalists have tried to defend naturalism against these kinds of philosophical anomalies by making the epistemic argument that naturalism is not required to produce explanations. Some beliefs are simply *basic* in the sense that they do not need any further justification, and so a naturalist does not need to engage in the broader epistemic task of explaining particular beliefs about reality. This form of "quietism" is, according to Macarthur, not successful and does not provide an escape route for the scientific naturalist. Contrary to this epistemic claim, Macarthur concludes that scientific naturalists, given their commitment to the continuity thesis, must be able to show how science might one day explain the kind of higher-level phenomena that seem indispensable in our human practices.

Liberal Naturalism and Religious Naturalism

Liberal naturalism constitutes an imaginative alternative to reductionist versions of naturalism. It seeks to account for higher-level properties, such as normative and mental properties, and abstract objects, without reducing such properties to the physical base level. Given a liberal naturalist view of reality, there might be some hope when it comes to connecting the manifest image with the scientific image of reality, and thus avoiding metaphysical grounding problems. Furthermore, given this view's ontological and methodological tolerance and the assent to causal pluralism, it also seems compatible with many of the indispensable presuppositions that we must make in various human practices. Liberal naturalism might provide a way forward for religious naturalism.

Both monistic naturalism and pluralistic naturalism are explicitly engaged in the business of explanation, and in the project of grounding higher-level phenomena in a naturalistic worldview. We could say that these two forms of naturalism engage in *positive* metaphysics, in the sense that they seek to positively find naturalistic ways of interpreting, explaining, or accommodating higher-level entities, albeit in different ways. Liberal naturalism instead defines the nature of naturalism *negatively*; an explanation must not contradict the laws of nature, or introduce entities that would in some way interrupt the causal nexus. Liberal naturalists such as De Caro, Macarthur, and Voltolini do not attempt to explain the existence of, for example, modal properties, or to specify the necessary conditions for their instantiation. It is enough, according to their view, to defend higher-level properties from reductionism, and to show that they do not bring with them anything metaphysically problematic that would threaten the idea of an unbroken causal web.

For many religious naturalists, the notion of mystery is essential for distinguishing generic (or purely secular) naturalism from their position. It is this sense of mystery in face of the richness of reality that gives rise to awe and wonder. By virtue of emphasizing the limits of knowledge and the unpredictability of emergent or higher-level properties, liberal naturalism is certainly compatible with and even reinforces the idea of the mysterious dimensions of physical reality. Liberal naturalism might then do the same kind of job as pluralistic naturalism, religiously speaking, with the significant benefit of avoiding the metaphysical challenge of having to ground emergent properties in the natural order.

Evaluating Liberal Naturalism

The liberal position provides an interesting direction for exploring the meaning of naturalism. The critique that De Caro, Voltolini, and Macarthur aim at scientific naturalism seems legitimate. However, liberal naturalism suffers from a number of significant philosophical problems:

1. Liberal naturalism is not able to justify its position when placed in competition with other ontological frameworks. This challenge is called "the problem of competing ontologies."

2. Liberal naturalism fails in being an economical explanation. This is the "epistemic simplicity objection."

3. The "naturalistic" aspect of this liberal position remains unclear. This objection is called "the vagueness of naturalism."

The Problem of Competing Ontologies

Thus far we have seen that liberal naturalists maintain that higher-level phenomena are irreducible to the categories of science, and that they are both ineliminable and ontologically independent. We have also seen that this more liberal version of naturalism constitutes a powerful metaphysical background against which one could construct a religiously inspired naturalism. Even though De Caro, Voltolini, and Macarthur have provided an interesting critique of scientific naturalism, it seems that the liberal naturalist position itself comes off as too weak. Why should we take the ineliminability, irreducibility, and ontological independence of higher-level properties to support naturalism, as opposed to its metaphysical rivals?

The theses of liberal naturalism rule out a few competing ontological frameworks, such as ontological eliminativism (the idea that higher-level phenomena can simply be eliminated). However, ineliminability (IE), irreducibility (IR), and ontological independence (OI) remain fully compatible with (and could be understood as supportive of) other nonnaturalist frameworks, such as platonic dualism, hyperdualism, panpsychism, and theism. There is no direct link between IE, IR, and OI such that the truth of IE, IR, and OI necessarily gives us a naturalist understanding of these properties. Liberal naturalism, given its ontological and methodological tolerance, indeed widens the realm of the natural, but perhaps to such an extent that this formulation of naturalism merely becomes one possible way to think about the existence of the noncausal properties to which De Caro, Voltolini, and Macarthur have appealed.

Of course, liberal naturalism would reject some nonnaturalist alternatives, such as supernaturalism, on the ground that it entails an ontological split in reality, and that nonnaturalist alternatives render higher-level phenomena unsolvable mysteries. However, even if this is true of supernaturalism, this still leaves us with panpsychism, panentheism, process theism, and emergent dualism. These ontologies can also add noncausal properties to their "list of existence." It is, then, not enough for a naturalist to point to certain properties and say that these properties can escape ontological reductionism; the naturalist must also *explain* why property X is a better fit for naturalism than other competing ontological frameworks.

Frank Jackson, who himself endorses a more reductive version of physicalism, stresses the need for naturalist explanations. He makes the distinction between *serious metaphysics* and *shopping list metaphysics*. For Jackson, "serious metaphysics" is the business of providing a true account of what the world is like, which is more than a simple shopping list. Serious metaphysics continually faces what I call "metaphysical grounding problems." A naturalist, even someone who adopts a more liberal approach, has to set some limits for what can exist, thus "serious metaphysics is the investigation of where these limits should be set."[7] Doing serious metaphysics is to also seek "comprehension in terms of a more or less limited number of ingredients, or anyway a smaller list than we started with."[8] In this way, the naturalistic metaphysicians seek a "comprehensive account of some subject-matter—the mind, the semantic, or, most ambitiously, everything—in terms of a limited number of more or less basic notions."[9] Following Jackson, it seems as if the liberal naturalist effort to maintain the reality of modal properties and other higher-level phenomena without explaining how such properties fit a naturalistic framework results in a metaphysical shopping list.

In a similar vein, Terry Horgan has argued that too little philosophical attention has been directed at what kinds of explanations might be possible with regard to naturalist understandings of supervenience. Horgan maintains that naturalistic explanations have either been too strict or too lax. According to Horgan, "It is not kosher to invoke supervenience relations unless they are subject to naturalistically acceptable modes of explanation."[10] Given that liberal naturalists do not provide naturalistic explanations for the reality of higher-level phenomena, in the sense of showing how a purely physical base level can produce high-level properties, they cannot be considered philosophically "kosher."

Naturalism, even in a more liberal version, has limited ontological resources. It is, I suggest, not enough to simply posit certain higher-level properties as irreducible or conclude that they are sui generis brute facts. I agree with Jackson and Horgan that we need some positive reason for thinking that they should be construed as specifically naturalistic properties. Otherwise, there seems to be an ontological split on the part of the naturalistic worldview: on one side, we have those properties that are definable in terms of complex arrangements on the base levels, but on the other side we have a variety of properties that seemingly escape any sort of explanation. This should not be an acceptable conclusion for a naturalist who seeks a unified view of reality.

The Epistemic Simplicity Objection

Naturalists, be they of a secular or a religious bent, take naturalism to be superior to other ontologies because of its apparent epistemic simplicity. Naturalism seeks to be continuous with the methods of science, as it pursues the simplest explanation so as to get rid of unnecessary entities; "the fewer entities the better," it is often said. In chapter 2, we could see that Loyal Rue and Willem Drees hold to this epistemic virtue. Quine expressed something like the simplicity principle in his philosophical explorations of naturalism. There is, he says, a spirit of reduction in science, and naturalists do their best to mirror the method of science and adopt the same spirit. Even if not explicitly stated, the idea here is that simplicity is a good indicator of truth.[11]

We have seen that liberal naturalists argue for the possibility of irreducible, unexplainable, and ontologically independent properties. In a way, De Caro, Voltolini, and Macarthur seem to reject the idea of epistemic simplicity. However, to connect this issue with the earlier objection, the threat of competing ontologies increases when one simply adds more properties to the list of things existing without providing an account of how higher-level properties relate to their physical base. Hence, there is a strong reason not to introduce unexplainable properties/entities to the naturalist framework and therefore to stick to the principle of epistemic simplicity. This is also related to Jackson's idea that serious metaphysics demands "comprehension in terms of a more or less limited number of ingredients."[12] Liberal naturalists clearly increase rather than decrease the list of ingredients.

The Vagueness of "Naturalism"

What kind of naturalism does this liberal perspective amount to? A scientistic version of physicalism is clearly ruled out from a liberal point of view. That is, the identity relation between higher-level properties and physical properties is not a viable option, given that there is no way to derive the former from the latter. It is important to point out that this is not merely an epistemic claim, given that liberal naturalists attribute ontological independence to higher-level properties.

Should we, perhaps, regard this liberal version as a form of nonreductive physicalism? There is no consensus on the exact definition of nonreductive physicalism, but it is usually formulated in terms of nonreducibility and supervenience. That is, Y is nonreducible to its physical base X, and any changes with regard to Y must correspond to a change in X.[13] From this we also get the saying "No mental difference without a physical difference." Supervenience is often regarded as a minimum requirement for a robust naturalism or physicalism.

Liberal naturalists and nonreductive physicalists both agree on the inadequacy of reductive physicalism, whereby such kind of physicalism is considered to lead to an identity relationship between mind and matter. Nevertheless, there is an important difference between these two views that makes me doubt that liberal naturalism should be construed as a type of nonreductive physicalism. Liberal naturalists further claim ontological independence for these extraphysical properties, meaning that they are ontologically independent of the entities describable by science. This ontological tolerance of liberal naturalism, however, seems to mean that we cannot hold to the often-assumed thesis of supervenience. Supervenience is a dependency thesis, and this does not go well with the claim that there are properties that exist "above and beyond" the physical entities that science deals with on a regular basis. Consequently, liberal naturalism is most likely not intended to be a version of nonreductive physicalism.

Another possible interpretation of liberal naturalism is that it is a version of emergentism, related to the pluralistic ontology that is proposed and argued for by some religious naturalists. Comparing the two, we find some important points of agreement: higher-level properties are nondeducible from physical stuff, and these "above-and-beyond" properties exist in an ontological and realist fashion. That is, they are neither solely epistemological constructions, nor are they reducible to a way of talking about reality.

However, when an emergentist claims that a variety of properties "arise from" their physical base, she is committed to some kind of supervenience, whereby a property is ontologically dependent on the microphysical structure. As Jaegwon Kim puts it, "Emergent and resultant properties of a whole supervene on, or arise out of, its microstructural, or micro-based, properties."[14] The expression "arise from" can very well be replaced with "supervene on," according to Kim. But, as pointed out, supervenience and liberal naturalism are not compatible; thus, this form of naturalism, if it is a version of emergence, cannot be the type of emergence that presupposes or entails the idea of supervenience.

Indeed, some prominent emergentists explicitly deny the need for supervenience. Timothy O'Connor, being one of them, has argued that we need to adopt a robust metaphysical conception of emergent properties in order to avoid the threat of possible reductionist explanations. He rejects the supervenience slogan that "there can be no mental difference without a physical difference." Moreover, argues O'Connor, an emergent state E1 can produce another emergent state E2 without there being any type of physical → mental correspondence. This is a very interesting take on emergence, and it deserves to be engaged with in a serious way. However, as others have pointed out, this view of emergence, suggesting a lack of causal correspondence on the physical level, seems to be closer to dualism than emergentism, at least with regard to how emergence has been conceived traditionally (for example, by the British emergentists).[15]

Another reason not to view liberal naturalism as an expression of emergence theory is that liberal naturalists reject the causal efficacy of higher-level properties, as they think that such causal efficiency would contradict the laws of nature. Naturalists of the liberal inclination, as we have seen, oppose the introduction of causes within the natural order that do not fit a view of reality as being governed by unbreakable natural laws. The notion of causal efficacy, or downward/top-down causation, is an essential part of emergence theory, either in a weaker form (that retains supervenience) or a strong one (that rejects supervenience). As suggested earlier, the reason that emergentists emphasize these kinds of higher-level causal capacities is to avoid the problem of epiphenomenalism.[16] Hence, proponents of liberal naturalism and emergence theorists differ substantially on the importance of the causal efficacy of higher-level properties.

We can then conclude that liberal naturalism should not be interpreted as a strong form of emergentism, for such an interpretation would imply that liberal naturalism is closer to dualism than naturalistic monism. Given that liberal naturalism clearly rejects dualism, and effectively construes naturalism as the negation of dualist ontologies, this cannot be the correct interpretation of this project. The liberal form of naturalism proposed by De Caro, Macarthur, and Voltolini is not a reductive version of naturalism, as it is clearly rejected. It is not a form of nonreductive physicalism, and neither can it be construed as a form of emergence. In the end, it remains unclear how we should understand the naturalistic aspect of liberal naturalism.

Religious Naturalism, Liberal Naturalism, and Agnosticism

To conclude this discussion, it seems quite difficult to pinpoint the naturalistic aspect of liberal naturalism. It fails to be a version of nonreductive physicalism, as it implies the rejection of supervenience. Nor does it seem to imply strong emergentism, given that De Caro, Voltolini, and Macarthur explicitly reject the causal efficacy of higher-level phenomena. Liberal naturalism should be undesirable to religious naturalists due to the vagueness of the liberal position. I have also suggested that liberal naturalism has a real problem with regard to the availability of "competing ontologies." Moreover, this ontologically and methodologically liberal approach seems to depart from the principle of epistemic simplicity, which historically speaking has been one of the defining features of naturalism. The liberal approach brings with it some philosophical benefits compared to pluralistic naturalism, yet religious naturalists should not "widen the realm of the natural" in the way that some liberal naturalists have recommended. I will instead investigate another possible way for moving religious naturalism forward. The direction I want to explore and analyze is called "agnostic naturalism." As we will see, and which has been explored in earlier chapters, agnosticism seems already to be a part of religious naturalism.

Agnostic Naturalism

In this section I will consider the agnostic version of naturalism, developed and defended by Colin McGinn, to see if it can provide an adequate background against which to view religious naturalism. First, though, I am going to outline the basic ideas and commitments of this perspective.

McGinn has argued that the body-mind problem is unsolvable. That there is a problem at all is, according to McGinn, obvious through simple introspection. He writes, "We know by just means of self-consciousness what the essence of consciousness is, and that is not caught by these kinds of attempt at specifying this essence. Our acquaintance-based concept of consciousness is not capturable in these attempts at knowledge by description. This is why we can sense a problem about consciousness just by introspecting it, without having yet articulated any of the properties listed: hence our natural, if excessively blunt question: 'How can *this* be *that*?' "[17] We do not have to provide a full list of the properties associated with consciousness to fully appreciate the philosophical problem of locating consciousness in the brain, or to think of mind as just a collection of neural processes. The problem arises quite naturally when we consider the boggling question of how brain states can produce conscious states.

According to McGinn, current attempts at explaining conscious activity all fall short and fail to understand the ontological distinctiveness of consciousness. Subjective experiences seem to go beyond mere causal descriptions and the language of neuroscience. He asks us to consider the appearance of pain. It seems as if how pain feels from the inside "is nothing like the way C-fibres appear when you look into a person's brain."[18] Hence, a version of naturalism that assumes an identity relation (a reductive relationship) between the physical and the mental should be rejected. Given the problems of explaining subjectivity through reductionism, property dualism has become the favored position among philosophers of mind. McGinn, however, maintains that property dualism is not an explanation: "In not telling us what mental properties are, it leaves them unconnected explanatorily to the brain."[19]

It is not just that contemporary solutions to the mind-body problem are explanatorily inadequate, and that a future solution is still possible (perhaps after the completion of the physical sciences). The problem runs much deeper. According to McGinn, this problem is unsolvable given that the real solution is unavailable to cognitive creatures such as we are. McGinn writes, "Minds are biological products like bodies, and like bodies they come in different shapes and sizes, more or less capacious, more or less suited to certain cognitive tasks."[20] Humans are simply not cognitively suited for solving the enigma of consciousness.

Given this cognitive closure of human beings, it is not surprising, McGinn goes on, that all constructive attempts to account for mind have failed and that there is little agreement to find in philosophy of mind pertaining to what kind of explanation to pursue. However, we should not simply raise our hands in despair and declare the emergence of mind as inherently miraculous. Neither should we proceed in an irrealist direction and dismiss the body-mind problem as illusory. Instead, McGinn proposes a nonconstructive solution, which means that we can appreciate a problem without being able to offer a solution to it. McGinn writes, "I think it is undeniable that it must be in virtue of *some* natural property of the brain that organisms are conscious. There just *has* to be some explanation for how brains subserve minds."[21] That is, there is some *natural* property *P* that can explain the origin of mind and the nature of the body-mind relation. There is, then, a naturalistic explanation, although we will never grasp the nature of *P* due to the inescapable limitation of our cognitive faculties.

At first McGinn thought that this skeptical diagnosis should not be applied to other philosophical questions. Later, however, he concluded "that this type of view is more widely applicable than seemed so at first sight."[22] An agnostic position, yet naturalistic in nature, is also justified when it comes to meaning, the self, intentionality, and knowledge. McGinn therefore proposes a nonconstructive approach with regard to several of the deepest philosophical questions, suggesting that there might be an explanation for these issues, but that it is beyond our cognitive abilities, even though the explanation has to be of a naturalistic kind.

Agnostic Naturalism and Religious Naturalism

We saw in chapters 2 and 3 how pluralistic naturalism, given its view on emergent properties, comes close to an agnostic position. As a consequence of saying that "we are epistemologically unable to deduce high-level entities/ properties from low-level physical laws," we can no longer make ontological judgments about higher-level properties. Hence, we cannot predict the direction of the evolutionary process or the biosphere as a whole.

Religious naturalists in adopting this pluralistic version of naturalism seem to express a form of agnosticism concerning the emergence of nature. However, there is an important difference between the agnosticism of the emergent pluralists and the one expressed by McGinn. According to pluralistic naturalism, emergent properties are "essentially unpredictable" (to quote Stuart Kauffman). That is, they are *ontologically* unpredictable, meaning that regardless of what kind of supercomputers we may have in the future, we will never be able to predict the kind of emergent properties/phenomena that will come into existence. I argued in chapter 3 that this kind of epistemology is devastating for ontological emergence. If one maintains that emergent properties are essentially unpredictable and irreducible, that we cannot explain how they came into being, then one is no longer allowed to say anything with confidence about the causal correspondence between higher-level features of reality and their physical base structure. Consequently, we have a problem with epiphenomenalism and we must conclude that the epistemology of emergence theory undermines ontological emergence.

McGinn, in contrast, seems to hold to epistemic agnosticism, according to which certain higher-level properties are unpredictable for *us* (given our present condition with the type of cognitive faculties that we are equipped with), but leaves open the possibility that other, more evolved, creatures might be able to explain the philosophical issues of consciousness, meaning, the self, and so on. Perhaps our cognitive faculties a few million years from now will have evolved in such a way that we then will be able to apprehend the nature of consciousness and other phenomena that currently resist our various explanatory endeavors. Or, in another scenario, we will finally have constructed one of Kauffman's supercomputers that can carry out the necessary computations and so do the job for us. Therefore, it is possible that McGinn's more limited agnosticism can provide a way forward for religious naturalists, especially with regard to the epistemic problem for pluralistic naturalism as outlined in earlier chapters.

In considering the possibility of an agnostic religious naturalism, two problems for McGinn's proposal arise: the problem of competing ontologies, and the self-refutation problem.

The Problem of Competing Ontologies for Agnostic Naturalism

McGinn has argued that a conceptual reduction of consciousness fails and that there is no way to reduce phenomenal experience, for example that of pain, to neurological firings in the brain. There is something above and beyond the categories and scope of science. The reason for this situation comes down to the fact, according to McGinn, that we are not cognitively suited for solving the problem of consciousness.

Despite the fact that there is no constructive naturalistic solution available (and, perhaps, never will be), McGinn still maintains that there is some natural property that can explain the relationship between mind and body. Hence, even though we may never be able to offer a positive naturalistic solution to some of the ultimate issues in science and philosophy, we should still insist on a completely natural ontology.

However, as in the case of liberal naturalism, McGinn's agnosticism seems to encounter the problem of competing ontologies. It is not enough to appeal to the failure of reductive explanations to establish the truth of naturalism, or to conclude that naturalism is the best available option. If it is the case that mental phenomena somehow escape reductive explanations, why should we take that to support naturalism as opposed to dualism or panpsychism? McGinn might reply that dualism (and panpsychism) is off the table, as it does not tell "us what mental properties are, it leaves them unconnected explanatorily to the brain." Naturalism should therefore be the default position for the honest enquirer. However, McGinn's nonconstructive naturalistic solution does not seem to make us any wiser; indeed, it can hardly be considered a "solution" at all. Whereas the dualist imagines an immaterial substance as the locus of phenomenal experience, McGinn postulates instead a mysterious *P*-property that we have no way of explaining. He even writes, "Perhaps *P* makes the relation between C-fibre firing and pain necessary or perhaps it does not: we are simply not equipped to know."[23] Explanatorily speaking, McGinn's nonconstructive naturalism does not help to move the discussion forward, and such an epistemic maneuver does not offer us any good reason for preferring a naturalistic framework over nonnaturalism for understanding the philosophical issues of consciousness, values, and meaning.

The Self-Refutation Problem of Agnostic Naturalism

A second issue arising from McGinn's position is that it ends up being self-refuting, because of the way that it places strict epistemic limitations on human cognition. In saying that metaphysical rivals, such as dualism and panpsychism, are unsuccessful in explaining consciousness, McGinn needs to be able to say something positive about the mind or pick out a feature of human mentality that such competing ontologies cannot explain. In saying that metaphysical rival *X* cannot explain mind is to already assume that a particular feature of the mind is in need of explanation. In so doing, McGinn is saying something positive about the mind, which his own perspective prohibits him from doing. By rejecting rival explanations of consciousness, McGinn is tacitly assuming a particular view of what it means to be a conscious creature, which betrays his agnostic stance and his emphasis on the limitations of our cognitive faculties.

McGinn suggests that philosophical agnosticism applies not only to mind but to self, values, meaning, and intentionality as well. Consequently, the epistemic authority of philosophy and the belief that good philosophical investigation can shed light on deep metaphysical issues are therefore undermined. On this view, philosophy is strongly constrained by the limitations of mind. Yet McGinn's own position is a product of philosophical analysis, and so he is relying on the very epistemological route that he deems problematic due to cognitive limitations. I am not claiming that McGinn's position is *logically* self-refuting, in the sense that there is a logical conflict between two propositions. Rather, it is self-refuting because McGinn's agnosticism undermines the reliability of his philosophical project.

Pragmatic Naturalism

Huw Price, whom I mentioned in chapter 3, has argued that the philosophical problems facing naturalism arise because we have conceptualized the ontology and epistemology of naturalism in a metaphysical way. His solution to these problems is to reconceptualize naturalism from a *pragmatic* perspective. The task of the naturalist, traditionally speaking, is to match ordinary discourse with the reality uncovered by science. The problem, as Price sees it, is that there seems to be a "striking mismatch between the rich world of ordinary discourse and the sparse world apparently described by science."[24] Due to this mismatch, naturalists have opted for different "placement strategies." The eliminativists faced with the excess of "higher-level" talk argue that "the statements we take to be true are actually systematically false."[25] The fictionalists hold instead that certain language games are "useful fictions" and that false claims can serve a useful purpose. The expressivists, in contrast, maintain that some statements are not genuine statements at all "but utterances with some other point or function."[26]

Price distinguishes between *object naturalism* and *subject naturalism*. Object naturalism is the type of naturalism we usually encounter in philosophical debates. This view states that "all there *is* is the world studied by science" and "all genuine knowledge is scientific knowledge."[27] Price suggests that it is object naturalism that gives rise to various placement problems. The task of the philosopher who holds to object naturalism is to match certain properties and different ways of talking with what science has revealed about reality. Price deems object naturalism unsuccessful in achieving this goal. Subject naturalism, by contrast, means that "philosophy needs to begin with what science tells us *about ourselves*. Science tells us that we humans are natural creatures, and if the claims and ambitions of philosophy conflict with this view, then philosophy needs to give way."[28] One could challenge this distinction and say that if you adopt object naturalism, then you are necessarily committed to subject naturalism, so what Price is offering to this discussion is both confusing and misleading. That is, if you accept that all real entities are natural entities, then you have no other option than to concede the naturalness of creatures like us. Price suggests that this negative conclusion is mistaken, as "subject naturalism is theoretically prior to object naturalism, because the latter depends on validation from a subject naturalist perspective."[29] Thus one is also free to respond to and critique object naturalism from the standpoint of subject naturalism.

Subject naturalism enables us, according to Price, to rethink the nature of the placement problem. For an object naturalist, the placement problem is material in nature; that is, we wonder how a certain object or property can also be a natural fact. The subject naturalist sees another alternative. The problem is *not* how to match nonnatural fact X with natural facts. Instead, we seek to understand how natural creatures like us came to use term X in a particular way. This takes us from a materialist to a linguistic conception of the problem. The problem facing naturalism pertaining to the oddity of higher-level discourse is fundamentally a problem of human linguistic behavior.

Price maintains that this shift from object naturalism to subject naturalism, and hence viewing naturalism as a collection of linguistic behaviors, allows for genuine pluralism. This form of pluralism considers seriously the variations between ordinary speakers. This is a kind of pluralism for which object naturalism seems unable to offer a satisfactory explanation. Price writes, "I think pluralism can do much to account for these variations, by appealing to the different functions or linguistic roles of the different domains concerned. And *only* pluralism has the flexibility to do this: other approaches are tied too rigidly to the existence of a fact-stating-non-fact-stating dichotomy in language."[30] In the end, Price suggests that subject naturalism, being more sensitive to the pluralistic nature of reality, is better equipped to harmonize our human practices with science. In this way, Price is led to adopt *global expressivism*, which rejects representationalism completely, and thus also the "project of theorizing about word-world relations."[31] Expressivism departs from realism and takes truth statements to have some other point or function. The expressivist approach therefore focuses on the behavioral dispositions of language users, and it "investigates the functions of the linguistic usage in question, and associated psychology, in non-representational terms, and then tries to account for the representational 'clothing.'"[32] In focusing on the "function" of language and the behavior of language users, Price moves naturalism in a pragmatic direction.

Pragmatic Naturalism and Religious Naturalism

I have argued with regard to monistic/pluralistic naturalism and liberal naturalism how the naturalistic project, thought of as a metaphysical doctrine about the nature of reality and the objects within it, creates a host of deep philosophical problems to which these frameworks seem incapable of providing answers. For a pragmatic naturalist who interprets the nature of naturalism linguistically rather than materialistically, the philosophical goal is not to reduce every higher-level property to the vocabulary of science. For Price, the naturalist should "abandon the project of theorizing about word-world relations" in a representationalist kind of way.[33] Price advises the naturalist to abandon object naturalism in favor of metaphysical quietism, similar to how some religious naturalists adopt quietism concerning God-talk. The issue to be investigated now is whether religious naturalists should extend their pragmatic conception of religious discourses and include naturalism within it. That is, several religious naturalists adopt a pragmatic conception of religious discourse. Should they then embrace pragmatism or nonrealism when it comes to the function of naturalistic discourses?

In chapter 5 I raised some concerns with regard to the pragmatic understanding of religion that has been offered by religious naturalists. Religious naturalists have not provided sufficient reasons for rejecting religious realism. Moreover, it does not seem obvious that whatever qualifies as an ecologically adequate image of God also qualifies as adequate from a scientific point of view, and vice versa. I further concluded, in chapter 3, that it is the metaphysical framework of naturalism that causes the most severe problems for the ongoing development of religious naturalism. Pragmatic naturalism might, then, be a promising alternative framework for a religiously inspired naturalist.

Evaluating Pragmatic Naturalism

Huw Price offers a fresh take on naturalism. His view of naturalism sets itself apart from traditional fictionalism/instrumentalism on the one hand, and reductionism on the other. Price should be commended for his honesty about the problem of squaring the numerous discourses of ordinary speakers with the kind of categories that are central to a naturalistic conception of science. Given the critique that I have aimed at both monistic and pluralistic naturalism, I suggest that Price is on the right track when seeking to develop a perspective above and beyond those that we typically come across in the naturalism-versus-nonnaturalism debate. He seems acutely aware of what I call the grounding problem for naturalism. The goal for Price is to defend higher-level talk while avoiding the metaphysics that typically accompanies such talk and that produces the hard problems of philosophy and theology.

Is Price's pragmatic take on naturalism the best framework for dealing with such pressing metaphysical problems? I suggest that we must answer in the negative for three different reasons: First, Price's linguistic conception fails to appreciate the seriousness of some philosophical problems. Second, global expressivism seems to undermine subject naturalism. And third, Price's pragmatic naturalism seems to imply realism. Consequently, pragmatic naturalism is not a satisfactory framework on which to base religious naturalism. I will expand on these objections below.

In the debate concerning the fate of the "M-worlds," that is, morality, modality, meaning, and the mental, naturalists affirm either eliminativism or some form of antireductionism. For Price, all these approaches fail, as they are stuck in the metaphysical matching game. The linguistic approach, which Price believed to be a forgotten rescue strategy, seeks instead to explain the M-worlds in terms of the linguistic behavior of ordinary speakers, thus showing how M-worlds are safe from reductionism. Linguistic naturalism is also called "functional pluralism": "A functional pluralism accepts that moral, modal, and meaning utterances are descriptive, fact-stating, truth-apt, cognitive, belief-expressing, or whatever. . . . Nevertheless, the pluralist insists that these descriptive utterances are functionally distinct from scientific descriptions of the natural world: they do a different job in language."[34] Price's linguistic treatment of philosophical problems is indeed interesting, and it might be a good candidate for some issues, such as modality. However, to explain, for example, mind and its capacities *solely* in terms of the agreement between certain speakers seems inadequate and quite reductionist. In the discussion regarding the mental realm we usually investigate questions such as "How can mental states have causal effects?," "Can mental states have effects in a causally closed universe?," "Are mental states physical or nonphysical?," "Is there an explanatory gap and, if so, can it be bridged?," and so on. To reduce these questions to the "speech habits" of ordinary speakers seems ill-founded. Moreover, Price is rather vague when it comes to specifying the functional tasks of different discourses. He says about various vocabularies (moral, scientific, or otherwise) that they might have a "multifunctional role in language" and "serve a range of very different functions."[35] But what these functions *are* remains unclear. Indeed, given that Price eschews all metaphysical talk, and seeks to take naturalism beyond representationalism, it is a challenge for Price to explicate what roles different vocabularies play. Michael Williams raises this concern as well: "But how are these roles to be characterized, if the language of philosophical anthropology must exclude any explanatory use of representationalist idioms?"[36] The problem for Price is not simply that he refuses to specify the variety of functions but that he cannot do so because such specification would itself be representational. He is stuck in the language game of traditional philosophy.

Applying pragmatism, and linguistic evaluation, on some areas might make sense. This would give us *local* expressivism. However, Price endorses *global* expressivism, and wants to extend pragmatism and the notion of linguistic priority so as to include all areas. Price maintains that semantics must stay away from the "descriptive task of providing ontological accounts of the relationship between the language and the world."[37] The dilemma for Price, then, is that due to his insistence on keeping a dividing line between semantics and ontology, he might not be able to explicate the functional roles of different vocabularies. The problem is that this entails not only metaphysical quietism but also, and more worrying, philosophical quietism.

If global expressivism invites philosophical quietism, then it seems that the pragmatism developed by Price severely undermines subject naturalism. Subject naturalism, according to Price, is the idea that we must start with what science tells us about ourselves—we are the philosophical starting point. And what science confirms in this story, according to Price, is that we are *natural creatures*. We do not have to invoke the more problematic framework of object naturalism. For a naturalist, subject naturalism should be sufficient. However, if Price is not just a local expressivist but, more strongly, a global expressivist, then subject naturalism and the language of science must be understood in an expressivist manner. According to this view, subject naturalism, like any other discourse, must be understood as an agreement between speakers. Price, however, rejects the idea that subject naturalism is merely the result of an agreement between speakers in a given discourse. Science reveals instead something ontologically fundamental about human nature, namely that we are natural creatures through and through, and that any philosophy that says otherwise must be rejected.[38] Price is then stuck between two, for him, unacceptable positions: either he takes his own global expressivism seriously and reduces subject naturalism to a linguistic agreement and concludes that subject naturalism is not truth-apt but that it serves a specific function; or he maintains that subject naturalism is more than a linguistic agreement, and so must reject global expressivism.

Price would, however, reject this conclusion, as he defends the normativity of truth. Truth, as he says, is not fiction, but *convenient friction*. This means that speakers understand there to be a norm of truth, that truth plays a certain role in dialogue. But is this not just plain antirealism or fictionalism? Price answers in the negative by suggesting that his own project completely sidesteps the "fictional-non-fictional" distinction and therefore cannot be accused of committing the errors of fictionalism. This makes it very hard to pinpoint the exact position of Price. Indeed, he says that his "account of truth is hard to find on contemporary maps."[39] However, given that subject naturalism is grounded in realist language (that this is, according to Price, what science really reveals about human nature), it seems as if Price's pragmatic and functionalist perspective remains, at least to an extent, stuck in realism. Hence, like other naturalists, Price is forced to tackle the grounding problem of naturalism—in this case, how to naturalize human nature completely. This would take us back to some of the metaphysical problems facing monistic naturalism and pluralistic naturalism. Price's naturalism, even though interesting, cannot offer the necessary resources for moving religious naturalism forward.

Conclusions

In this chapter I have investigated three alternative naturalistic frameworks, or ontologies, in order to see if any of these can provide the necessary aid for the project of religious naturalism: liberal naturalism, agnostic naturalism and pragmatic naturalism. They all attempt in different ways to avoid some of the classical grounding problems facing a naturalistic ontology. Liberal naturalism offers a sophisticated defense of several irreducible phenomena. However, it could not be considered successful, as it was difficult to discern the "naturalistic" aspect of this perspective. It was, moreover, unclear on this view why we should adopt naturalism rather than any of its metaphysical rivals.

Colin McGinn's agnostic naturalism was also discussed and evaluated. I highlighted the fact that several religious naturalists, in adopting emergence theory, express agnostic tendencies. However, as in the case of liberal naturalism, McGinn failed to justify choosing naturalism over other competing ontologies.

We have seen that several religious naturalists express a pragmatic understanding regarding the religious sphere. In this chapter we considered

Huw Price's pragmatic naturalism, and if a pragmatic understanding should include naturalism itself. I argued that Price's position is philosophically inconsistent and cannot be considered a viable alternative framework for religious naturalism.

7

Alternative Ontology 2

Theistic Options

In this chapter I will discuss whether religious naturalism should consider some versions of theism due to the problems of (monistic and pluralistic) naturalism and the various ways of understanding religious discourse (realistic, antirealistic, and pragmatic accounts). I will consider two versions of theism in this chapter: panentheism and Fiona Ellis's theistic proposal.

In the introductory chapter I briefly discussed panentheism. It was seen that panentheism shares several central beliefs with religious naturalism; for example, the belief that science undermines supernaturalistic causation and divine interventions in physical reality, and consequently that traditional religion must be modified due to the advancements and increasing knowledge of modern science. Furthermore, both panentheism and religious naturalism reflect a theological shift toward emphasizing divine immanence. However, I argued that panentheism is distinct from religious naturalism given that it affirms the ontological reality of God, that is, a reality separate from the physical world.

This chapter will explore the panentheistic framework to see if it is a viable option for religious naturalists. In the previous chapter I explored three alternative formulations of naturalism: liberal naturalism, agnostic naturalism, and pragmatic naturalism. It was seen that these naturalistic frameworks cannot help religious naturalism in moving forward, due to several severe philosophical problems and inconsistencies. Having explored a few alternative naturalistic frameworks, it is now time to explore some theistic alternatives.

This chapter proceeds in the following way. First, I situate panentheism in the dialogue between science and religion, and outline my argument. Second, I describe in what sense panentheism is a naturalistic position. Third, I will critically examine the dualistic implications invited by the notion of "psychological miracles," and the ontology of emergence theory employed by some panentheists. Fourth, I look at process panentheism and how it too is unable to avoid dualism, despite claiming to be a robustly naturalistic position. Fifth, I argue that panentheism suffers under Ockham's razor.

After the critical evaluation of panentheism, I will move on to consider Fiona Ellis's theological project, which seeks to synthesize Christian theism with John McDowell's expansive naturalism. I will argue that McDowell's naturalism, in a similar way to liberal naturalism and agnostic naturalism, encounters "the problem of competing ontologies." Toward the end of this chapter, I question whether Ellis's theistic grounding of values would be acceptable to a badge-wearing naturalist.

Background to Panentheism

Several thinkers in the science-religion dialogue maintain that classical theism is no longer tenable, and that science mediates a view of reality not compatible with the ontology expressed by a traditional or supernaturalistic theism. According to this view, the ontological gap between God and the universe is deeply problematic, since, in order to bring about a certain event, God has to break the natural laws governing the universe. This not only undermines the practice of science but may also challenge the rationality of God (who created these laws in the first place). Thus it is argued that we must find ways of conceptualizing God that are compatible with the scientific view of reality as causally closed.

In recent years a position commonly referred to as "panentheism" has become increasingly popular, and several scholars involved in the dialogue between science and religion argue that this form of theism may be a solution to the problem of divine action as stated above. Panentheism typically refers to the idea that nature is situated within the divine life. It is also said that creation is "the body" of God. Several panentheists also refer to their view as "naturalistic theism" or "theistic naturalism"—that is, a combination of theism and naturalism.

In this chapter I will argue that panentheism is unable to avoid ontological dualism. More specifically, it will become obvious that a form of dualism is still assumed with regard to how God is revealed in the mental world of humans. Furthermore, the naturalistic assumption underlying the panentheistic project risks making notions of divine influence ontologically superfluous. In this way, a panentheistic view of divine causality becomes vulnerable to the principle of Ockham's razor.

I propose that panentheists not only fail to uphold a naturalistic conception of divine action but also that, in light of these problems, they would be wise to reject the metaphysics of naturalism. However, if it is the case that panentheism fails in upholding a naturalistic conception of physical reality, then it seems as if this perspective is not an option for the religious naturalist.

Going Beyond Reductionism: Finding a Plausible Naturalism

The panentheists discussed within this chapter claim to subscribe to some form of naturalism, a naturalism that is perceived to be able to avoid the problems easily invited by reductionism, materialism, or eliminativism. One problem, according to these thinkers, is that naturalism, by being materialistic in its ontology, has implied an atheistic worldview or perhaps even a nihilistic view of reality. Thus many people of religious faith have by default rejected the naturalistic project. Panentheists seek to overcome the dichotomy between theism and naturalism and by doing so achieve progress in the dialogue between science and religion. This ambition, as we have seen, is also expressed by religious naturalists.

The process philosopher and theologian David Ray Griffin argues that in associating science with materialism, scientific naturalism has been distorted. Thus "science will be regarded as antithetical to any significantly religious outlook."[1] Scientific naturalism, according to Griffin, can be understood in a maximal or a minimal sense. Maximally speaking, scientific naturalism is equated with reductionism, determinism, atheism, and materialism. Science, according to this view, rules out human freedom, divine influence in the world, and any ultimate meaning in life.[2] Minimal naturalism, in contrast, simply means that supernaturalism, or supernatural causation, is ruled out, that "the world's most fundamental causal principles are never interrupted."[3] Naturalism in its minimal form is, according to Griffin, fully consistent with the "most fundamental assumption of the contemporary scientific worldview," and it does not in any way threaten a serious religious outlook on life.[4]

In a similar vein, Philip Clayton argues that some kind of naturalistic presumption has to be made in theology and science. This presumption basically means that "for any event in the natural order, . . . its cause is a natural one as opposed to a supernatural one."[5] There is, according to Griffin and Clayton, no room for supernatural causation, meaning that there is no dualism within the natural order, from a scientific perspective. Arthur Peacocke concurs with Griffin's distinction between maximal and minimal naturalism and writes, "The only dualism which such a [naturalistic] stance accepts is indeed that between God and all-that-is, the 'world'; it rejects any dualisms within the natural order itself, including humanity."[6] The only processes we have in this world are natural. There are no spiritual or supernatural laws operating within the natural order.[7]

Both Clayton and Peacocke subscribe to an emergentist ontology, according to which reality is hierarchically ordered into higher and lower levels. They believe that this ontology is able to avoid the problems associated with reductionism and eliminativism. Based on emergence theory, Philip Clayton has proposed the *panentheistic analogy*: "Just as human consciousness (mental properties and their causal effects) can lead to changes in the physical world, so also a divine agent could bring about changes in the physical world—if this agent were related to the world in a way analogous to the relationship of our 'minds' to our bodies."[8] Clayton and Peacocke therefore suggest that panentheism, in virtue of incorporating an emergentistic ontology, is compatible with a naturalistic outlook on reality. Thus this form of naturalism does not imply reductionism.

A final view to consider is presented by Mark Johnston, who reflects on the distinction and relationship between methodological and ontological naturalism. Some, in order to construct and justify a naturalistic ontology, point at the success of science. However, as Johnston notes, science shows both success *and* failure, and there is no easy way to derive ontological conclusions from science.[9] Is Johnston then satisfied with methodological naturalism? He seems reluctant to embrace ontological naturalism even though he would like it to be true given that it would help us to finally be insulated from supernatural intervention.[10] However, he later writes that process panentheism reveals a view of God that "is in no way at odds with the form of the natural realm disclosed by science: that is, a causal realm closed under natural law."[11] Thus despite his earlier hesitation, science, in the end, seems to support at least a minimal naturalistic view of physical reality as causally closed.

As we can see, the naturalism employed by Griffin, Peacocke, Clayton, and Johnston is causally defined and assumes some philosophical idea of causal closure. Of course, what it means for reality to be causally closed can be understood in a variety of ways.[12] A strong form of causal closure, as expressed by the physicalist David Papineau, would give ontological and epistemological priority to physics. The way that Papineau understands the causal closure of the physical brings it closer to a more reductive kind of physicalism.[13] Causal closure, in its weak form, as expressed by Jaegwon Kim, merely suggests that "if we trace the causal ancestry of a physical event we need never . . . go outside the physical domain."[14] That is, all causes are to be located within the natural domain, and there is no need to appeal to extranatural categories when explaining physical phenomena. This latter view seems to be expressed by the panentheists discussed in this chapter.

Panentheists, in trying to find an adequate definition of naturalism, are careful to avoid physicalist, reductionist, and eliminativist notions of naturalism, which would either eliminate God completely from reality or at least invite a deistic conception of divine action. However, they are also committed to finding a naturalism that can defend reality from supernaturalist interventions. A middle path is needed, according to these thinkers.

It is helpful here to distinguish between two forms of naturalism: global and local. A global naturalist would exclude all supernatural beings, realities, and categories. Global naturalism is a thesis about reality as a whole. A local naturalist, in contrast, would argue that we need only to adopt a naturalistic ontology with regard to a specific phenomenon within the world—in this case, causal processes. The panentheists in this chapter thus hold to local naturalism as opposed to global naturalism, as they maintain robust theism. This is an important difference between panentheists and religious naturalists, as the latter group maintains the stronger thesis of global naturalism.

We will next see if panentheists manage to uphold this middle path between dualism and reductive naturalism while saying that God actively engages with humanity and the universe.

Why Panentheism Entails Dualism

In rejecting supernaturalism, panentheists do not want to rule out divine activity altogether. Instead, they want to find an adequate model for construing divine activity so as to make it compatible with a minimal or local naturalist view of the universe as being governed by unbreakable natural laws.

As described above, Peacocke and Clayton argue that emergence theory shows how God might act in the natural world in a way that is consistent with a core commitment of naturalism. Emergence theory, according to Peacocke and Clayton, offers a view of divine causality that does not introduce the dualism associated with supernaturalism. A panentheistic conception of God's activity does not merely allow a general form of divine action, meaning that God sustains the universe and keeps it in existence. For proponents such as Peacocke and Clayton, emergence theory enables the panentheist to retain special divine action as well, such that God can accomplish particular purposes within the physical order, compatible with the broader philosophical assumptions of a naturalistic worldview.

Yet it seems that a form of dualism is still present in this emergentistic conception of divine action. Clayton, for example, does not rule out all forms of miraculous events, as he claims that psychological miracles are still scientifically possible. Given the emergentistic account of mind, he argues that the sciences have "not been able to formulate and corroborate fundamental laws of human behavior."[15] Thus one cannot rule out the possibility of God performing psychological miracles and speaking to people directly—clear instances of special divine action. One could, however, object to Clayton's idea of psychological miracles by saying that the future might yield new scientific discoveries confirming a type-type relationship between mental and physical states; hence there would no longer be any reason to talk about miracles in the mental realm. Clayton is aware of this objection but states that this reductionist program, in light of modern cognitive science, does not seem very promising. On the contrary, he avers that we have positive reasons for rejecting reductionism or eliminativism and to affirm the irreducibility of mental states.[16]

Here it becomes clear that Clayton does not stay true to his naturalistic and antidualistic commitment. There is, on his view, undeniably a form of dualistic causation, causes within the mental realm that are not reducible to something physical and that transcend natural laws. We have no lawlike explanation for these events, and thus it is possible, even today, to talk about special divine action within the mental life of human beings. God, according to Clayton, performs psychological miracles such that new mental content is inserted into the mind of the religious believer, and this content cannot be reduced to mere neurophysiological happenings in the brain. This, however, contradicts the presumption of naturalism, which claims that for every event (including mental events) there is a natural cause, as opposed to a supernatural or extranatural one. Clayton seems to think that the mental realm is God's last stronghold, and if we reject the notion of divine mental influence, then we end up with deism. I think that his worry about deism is justified. If the reductionist program is successful, then, on

Clayton's view, there would be no room left for God within the natural order and so deism would be the only available theistic option. But in his attempt to avoid deism, Clayton has adopted a dualistic causality because there are causes in the world that transcend the limits of natural science, that are ontologically irreducible. Mental divine revelation, or what he calls psychological miracles, invites dualism rather than naturalism, given that their causal origin is not physical; it is to be located in the mind of God.

Peacocke's perspective, it seems, faces the same dilemma. He suggests that in his emergentist model, whereby God acts in the world through whole-part and top-down causation, providential action is possible. He writes that "*particular* events or clusters of events, whether natural, individual, and personal, or social and historical . . . can be intentionally and specifically brought about by the interaction of God with the world."[17] By making room for special divine action in the world, Peacocke seeks to avoid a deistic conception of the God-world relationship. However, at this point a dualistic ontology seems to emerge.

If Peacocke maintains that God works purposefully in his interaction with the world, then there seems to be a difference between the specific acts of God, and ordinary, nonspecific/nonintentional happenings in the natural order. Some happenings have their causal origin in the intentionality of God; some do not. Here we can detect an ontological difference, which will be discussed further below. If Peacocke denies that there is any difference between God's providential acts and the ordinary happenings in the world, then there is no longer any reason to talk about *special* providence. They would become identical, and, as Clayton puts it, there would no longer be any "reason to interpret it [an effect in the world] as an instance of divine action."[18] For it to be meaningful to talk about "divine action," there needs to be some ontological difference between God's actions and natural events. Thus special providence, as formulated by Peacocke, seems to invite dualism, and this entailment undermines minimal or local naturalism and the philosophical principle of causal closure.

Things get even trickier when we consider the metaphysical implications of emergence theory. Both Clayton and Peacocke construe panentheism, God's interaction with the world, against the background of emergence, and they understand emergence theory to be compatible with metaphysical monism. In a similar vein to religious naturalists, the panentheists discussed here claim to subscribe to some form of monism. This form of monism, Clayton explains, means that "there is one natural world made, if you will, out of stuff."[19] Similarly, Peacocke writes that he prefers to call his position emergent monism, rather than nonreductive physicalism: "This is a 'monistic' view that everything can be broken down into fundamental physical entities and that no extra entities are thought to be inserted at higher levels of complexity to account for their properties."[20] In essence, their view states that there is only one natural world, although nature is layered.

However, as in the case of religious naturalists, Clayton and Peacocke encounter the problem of how to square the notion of downward causation with the monistic idea of causal closure. Quite similar to religious naturalists, Clayton and Peacocke subscribe to strong emergence, the defining feature being the concept of downward causation. But, as I argued in chapter 3, downward causation, when more closely scrutinized, does not seem to be compatible with causal closure; it in fact supports dualism.

This dualistic conclusion becomes obvious when we discuss divine action in relation to panentheism. Certain events will have their causal origin in God. But if it is the case that certain phenomena or events are dependent on or have their causal origin in a nonnatural reality, then this view is not consistent with naturalism and the idea of causal closure. In this way, a theist who aims to explicate how God interacts with reality via emergence theory would invite a dualistic conception of causal processes, not a naturalistic one. Of course, if my argument is correct, then it is downward causation that gets emergence theory into trouble. One option for the panentheist then would be to abandon downward causation and merely hold to weak emergence (the idea that reality is dividable into higher and lower levels). The notion of downward/top-down causation, even though controversial, is crucial to emergence theory, as it is easier to talk about the ontological realness of emergent property X if X makes a causal difference. Peacocke writes, "For *to be real is to have causal power*. New causal powers and properties can then properly be said to have *emerged* when this is so."[21] By getting rid of strong emergence, a panentheist might be able to avoid the dualist implications of emergence theory. But if Peacocke is correct, then downward causation is an ontologically necessary feature for talking about the realness of emergent properties/phenomena, and therefore for avoiding reductionism.

Recall also my argument (in chapter 3) regarding the problematic consequence of the unpredictability thesis of emergence theory, which was shortened to *PON3* and *PON4*.[22] The idea of emergent properties as being essentially unpredictable leads to ontological agnosticism. Thus the idea that we cannot predict emergent phenomena makes it impossible to offer a positive account of the relationship between higher and lower levels of reality. The threat of epiphenomenalism increases, therefore, for the emergence-committed panentheist. Emergence theory, as expressed by Peacocke and Clayton, leaves us with a form of mysterianism. Emergence theorists may have provided good arguments against reductionism; nevertheless, given the problematic task of providing a positive account of emergent phenomena, it seems unwise to use this theory when trying to positively explicate the relationship between divine causality and the natural order. Panentheists, just like religious naturalists, encounter severe philosophical problems in their use of emergence theory. Therefore, emergent panentheism seems unable to help religious naturalism to move forward.

Process Panentheism and the Dualistic Implications of Divine Mental Revelation

The process panentheism proposed by Griffin and Johnston seems to run into similar problems. However, contrary to emergent panentheism, Griffin denies the doctrine of *creatio ex nihilo*. Thus he takes one step further in an antidualist direction. Perhaps process panentheism, for that reason, is more successful when it comes to avoiding both global and local dualism, and thus is even more interesting for a religious naturalist. The reason process panentheists deny creation *ex nihilo* is that it inevitably invites supernaturalism and divine interventionism. The doctrine of creation *ex nihilo*, according to which God created the world out of absolutely nothing, implies "that the world has no inherent power, no power of its own, with which it could resist the divine will."[23] Given creation *ex nihilo*, there are no autonomous natural laws. Process panentheists believe that classical theists are unable to uphold the view of natural laws as autonomous. Classical theists can instead refer to natural laws only as God's usual way of treating the natural order. Therefore, there is nothing stopping God from suspending these "laws" and violating the integrity of creation.

Has Griffin, then, managed to eradicate the last traces of dualism? It seems that he has not. Griffin is very clear about one thing: his form of naturalistic theism is not to be confused with naturalistic approaches to religion as exemplified by Paul Tillich, John Dewey, and religious naturalist Gordon Kaufman. Their error, according to Griffin, was to completely collapse the distinction between God and the world. That is, they proposed pantheism, according to Griffin. There is a sense, says Griffin, in "which process theism could intelligibly be considered a type of supernaturalism."[24] It is supernaturalistic in the sense that (contrary to a pantheistic position) there is an ontological difference between God and physical reality, although physical reality is contained within the divine life. Thus Griffin, in order to avoid pantheism (which he considers to be a form of atheism), upholds a dualistic distinction between God and the universe.

What, then, shall we say about Griffin's and Johnston's idea that process panentheism is committed to a nondualistic view of divine influence? Johnston claims that "God reveals himself in the natural realm by disclosing in religious experience an ultimate form of the world, one that is in no way at odds with the form of the natural realm disclosed by science."[25]

Griffin argues that we must perceive God's influence in the natural order in terms of persuasion rather than coercion, since the latter notion of divine causation ultimately leads to God interrupting the basic causal-effect relations. Whatever God is doing in the natural order is done purely through natural causes. For example, Jesus's incarnation was fulfilled solely through persuasion, meaning that God was present in him "in such a way as to reveal God's character, purpose and mode of operation."[26] God was incarnated in Jesus in a special way, and we can therefore rightly affirm that Jesus is God's decisive revelation without invoking supernatural causation. Jesus was not inherently the incarnate Son of God, in a supernatural sense, but became God's revelation through God's persuasion of the man Jesus.

However, the same problem of dualism returns again. If God, as Johnston and Griffin claim, is revealed through religious experience, and that God introduced new content in the minds of religious believers, then God obviously made a causal difference in reality. Moreover, the reason that S has a religious experience, why God has inserted new mental content in reality, cannot ultimately be explained in terms of the physical structure of the brain or by reference to natural laws. The causal origin of God's revelation (the new mental content disclosed by God) is to be traced back to divine intentionality, not to some physical happening in the brain. God may, on Griffin's and Johnston's view, use the physical structure of the brain to produce these religious experiences, but they cannot be reduced to neurochemical events. This is because God directly affected the brain structure, bringing about new mental content (and therefore intervened in physical reality). The causal closure has been broken; local naturalism is violated.

Panentheistic Divine Action and the Threat of Ockham's Razor

We should be absolutely frank here about the philosophical implications so far: either God makes a causal difference or God does not. If you maintain that God does make a causal difference, then extranatural influences are operating within the natural order. Hence, you cannot consider yourself to be a naturalist—not even in a minimal sense. But if you deny the causal relevance of God and say that there is no ontological difference between God's actions/divine influence and ordinary causes, then there is no longer any reason to talk about divine action or divine influence. Consequently, God is expelled from the natural order, and thus you end up with deism. The middle path between classical theism and deism that panentheists strive to defend is, one should say, philosophically unstable.

A panentheist would presumably respond by saying that there is a difference, but that there is no *qualitative* difference, that is, no ontological otherness, between God's influence and the regularity of natural law. Clayton himself holds to this view.[27] However, Clayton also writes that "when someone claims that God accomplished a particular action, she is saying that the event cannot be explained in the way in which we normally have knowledge of natural events."[28] How should we interpret this? Given that Clayton denies that there is any ontological difference between God's influence and natural regularities, two ways of interpretation remain: (a) semantically: higher-level vocabulary about claimed divine occurrences cannot be explicated in terms of physical language and the language we use in reference to natural laws; and (b) epistemologically: claimed instances of divine action cannot be explained by science. We cannot, and we will never be able to, explain claimed divine occurrences due to the limitations of science.

If we want to be able to say that God makes a difference in reality, then it seems that (a) and (b) are too weak. The former amounts to a semantic thesis that is fully compatible with ontological reductionism—roughly, the idea that the only real processes in the world are those taking place on a microphysical level.

The latter claim can be interpreted in two different ways, one weaker and one stronger. The weaker version would suggest that we *currently* cannot explain claimed divine occurrences from a scientific perspective. One could also maintain something stronger and say that scientists will *never* be able to explain these claimed divine occurrences. If true, (b), expressed in either a weak or a strong form, will establish only epistemological irreducibility. However, epistemological irreducibility, just like semantic irreducibility, is consistent with an ontologically reductionist view of the world. It might be the case that we will never arrive at a successful and purely physical account of claimed divine occurrences, even though the only real processes in the natural order are mindless, nonteleological physical/material processes. The epistemological limitation of science is not sufficient to safeguard divine influence from the reductionist spell.

It is the commitment to local naturalism, and the philosophical thesis of causal closure, that prohibit panentheists from saying that there is an ontological difference between divine influence and natural causes. This philosophical commitment, however, runs the risk of making the notion of divine influence ontologically superfluous. If there is no ontological difference between God's action and the regularity of nature, why introduce God-talk in the first place? God-talk, in panentheism, does not in any way help to explain the structure of reality; it is an unnecessary metaphysical add-on. This offered panentheistic notion of divine influence becomes vulnerable to the principle of Ockham's razor. This principle states that we should, among competing hypotheses, adopt the one that contains the fewest assumptions. That is, a hypothesis that posits unnecessary entities, or that needlessly introduces ontological categories, should be rejected in favor of a simpler one. A panentheistic notion of divine influence, by denying any ontological difference between God's action and natural causes, becomes ontologically superfluous and, by virtue of Ockham's razor, should be rejected. Panentheists must choose. Either they affirm that there is an ontological difference between God's actions and natural regularities (thus introducing a form of causal dualism), or they collapse the distinction between the two, which means that divine action, as imagined by panentheists, is vulnerable to Ockham's razor.

The Positive Status of Panentheism and the Question of Religious Naturalism

A main argument of this chapter has been that panentheists such as Clayton, Peacocke, Griffin, and Johnston should reject the idea that the universe is causally closed. Indeed, if my argument is correct, that the ontology of strong emergence implies dualism, then emergent panentheists in particular should reject local naturalism.

In order to avoid making notions of divine influence ontologically superfluous, panentheists need to rethink the nature of divine action and finally leave local naturalism behind. God must be allowed to make a causal difference in reality, and for this to be possible there needs to be some qualitative/ontological difference between God's influence and natural laws. Of course, this would imply that, from a theistic perspective, scientific causal explanations will be incomplete. As Keith Ward writes, "If

God's intentions plus the physical laws do give a sufficient explanation of what happens in the world, whereas the physical laws alone do not, it is clear that the existence of God contributes substantially to an explanation of what happens in the world."[29]

Thus far I have argued that panentheists are not consistent in that they still assume a dualistic ontology with respect to causal processes while claiming to adopt local naturalism. That is, they *cannot* hold to local naturalism while claiming that God actively influences the natural order. However, given the discussion above regarding the issue of Ockham's razor, I also want to emphasize that panentheists *should not* adopt local naturalism, as this commitment undermines the ontological relevance of ascribing divine importance to particular events. The idea of causal closure, an important commitment for several panentheists, clashes with the ambition to uphold a strong notion of divine causation. Therefore, the forms of panentheism discussed in this chapter cannot be considered viable alternative frameworks for religious naturalism. Because they stress the active role of God in the unfolding of creation and the future of humanity, the forms of panentheism that I have analyzed fail to avoid dualism and cannot be considered naturalistic in any meaningful sense of the term. Religious naturalists must look elsewhere.

Next, I will consider Fiona Ellis's ambition to combine Christian theism and John McDowell's naturalism.

God and Values: A Proposal by Fiona Ellis

Fiona Ellis argues that contemporary understandings of naturalism carry an unnecessary commitment to scientism, according to which science is the only path to knowledge. Given that the question of the existence of God lies beyond the reach of science, God's existence is either denied, or the question is deemed meaningless. Consequently, Ellis argues that if theism is going to be a real option, we must find ways of conceptualizing "naturalism" that takes us past "scientific naturalism" with its strong commitment to scientism.

Ellis's project consists of two tasks. The first task is to evaluate different forms of nonscientific naturalism to see if any of these ontologies can account for values. The second task is to show that how we relate to "value is both necessary and sufficient for relating to God."[30] In a similar way to the different naturalistic proposals examined so far (monistic, pluralistic, liberal, agnostic, and pragmatic naturalism), Ellis seeks to go beyond a narrowly defined scientific and reductionist version of naturalism. This evaluation of Ellis's proposal will proceed as follows: I will describe Ellis's critique of reductionist formulations of naturalism and her endorsement of John McDowell's nonreductionism. I will thereafter describe and evaluate Ellis's attempt to fuse a robust Christian theism with McDowell's naturalism.

Beyond Scientific Naturalism and Supernaturalism

Naturalism is a difficult concept to grasp, and many have suggested that naturalism is no longer a distinctive "–ism." Ellis is aware of the fluidity of "naturalism" but suggests that it can be both negatively and positively defined. At the negative level, the naturalist rejects supernaturalism and the idea of beings or realities transcending the natural order.[31] Positively, the naturalistic position is characterized by a strong commitment to the scientific method. It is because of this latter commitment that theism has been considered to be incompatible with naturalism. Hence, "we are left with the claim that supernaturalism is to be rejected because it allows that there is more to reality than what the scientist can comprehend."[32] Referring to, among others, Mario De Caro and David Macarthur (whom I discussed in the previous chapter), Ellis suggests that a naturalism that elevates science over all other disciplines (and that therefore seeks to reduce all other disciplines to the natural sciences) is problematic, as such a view cannot be supported by science. Moreover, to suggest that we can explain all of nature and human behavior in terms of purely physical stuff is to attribute "supernatural powers" to the physical level, and this goes contrary to the naturalistic position.[33] In order to understand the natural world, we need to invite more disciplines to join this epistemological endeavor. Scientific naturalism is too strict for its own good and cannot account for values. We need a liberalized version of naturalism, an "expansive naturalism."

God and Expansive Naturalism

The first version of expansive naturalism that Ellis considers is "expansive scientific naturalism." This naturalism "grants that there is more to science than physics, and more to it than natural science."[34] This naturalism is willing to include the human sciences within its methodology, and it rejects the epistemological monopoly of the natural sciences. On my typology, it seems as if a scientific expansive naturalist holds to *MON1–4* (that all of reality is made up of the same constituents, that there is a level of determination between higher and lower levels) and *PEN1–2* (that there are other disciplines capable of producing knowledge about reality, and that many meaningful claims fall outside the boundary of science).

However, despite being more open to an input from the humanities, "scientific expansive naturalism" cannot find a place for values. According to this version of naturalism, "human values are those which are valued by human beings."[35] Thus values are to be defined in terms of human interests. Ellis is skeptical toward such naturalism, suggesting that it cannot account for the normativity or "oughtness" of values. It can, as Ellis points out, be in someone's interest to be cruel.[36] Moreover, this kind of expansive naturalism, appealing to our interests, does not succeed in spelling out in detail the nature of values. There is an objective moral dimension that is missing.

Peter Railton, a proponent of expansive scientific naturalism, is aware of the problem of defining values in terms of human interests. He argues instead that values and moral norms should not be reduced to the opinions of individual persons but should be judged from a social point of view. Railton seeks to maintain a realist view of moral discourse and practice, and avoid a relativist view of values, by introducing some independent standard of judgment. Railton takes values to be reducible to "complex social-psychological phenomena," and the "aim is to show how the relevant phenomena 'may bring into being in individuals a notion of obligatoriness that will present itself to them as objective and independent of their personal inclinations.'"[37] In claiming so, Railton hopes to show how values, while being real, are not "spooky," "intolerably odd," or somehow ontologically different from other phenomena in nature. Ellis applauds Railton's ambition to retain a realist view of values. Yet Railton's approach comes off as too weak. It has the negative effect, argues Ellis, of squeezing moral properties out of the picture by reducing such properties to social-psychological phenomena. Railton is therefore still on the reductionist side of the spectrum, as he merely chooses to reduce values to *something else*.[38]

Ellis moves on to consider John McDowell's version of naturalism. Like other expansive naturalists, McDowell seeks to "defend the idea that moral properties are *sui generis*," by demonstrating the objectivity of values and showing that values are "part of the fabric of the world."[39] For McDowell, values are an inherent part of an enchanted nature that transcends the limits of science but does not contradict its methods or findings. Our practices, according to this view, are "already irreducibly value-involving."[40] In this way, McDowell's naturalism is antiscientistic (by denying that science has monopoly on ontology), but not antiscientific (as it still respects the insights gained from science). However, by saying that values are irreducible to the physical, we are not, according to McDowell, inviting supernaturalism or dualism. Our capacity to respond to the domain of values is a rational capacity that stems from our human nature, rather "than from any supernatural addition which gives us a mysterious separate involvement in an extra-natural world."[41] Our capacity to engage with values therefore requires nothing extrahuman, only our human nature. This capacity for recognizing values is fully natural, and thus consistent with naturalism.

Nature is in a sense enchanted, according to Ellis, who endorses McDowell's expansive naturalism. McDowell's naturalism takes us beyond a scientistic conception of nature, and toward a view of nature as infused with values. McDowell concludes that "there is more to explanation than scientific explanation, more to nature than what the scientist comprehends, and that once we concede these points we arrive at a partially re-enchanted conception of nature."[42]

Ellis bravely wonders if we should stop here and say that nature is enchanted with values, or if we could take this idea further and say that nature is *divinely* enchanted. An expansive naturalist, like McDowell, would be skeptical about the "possibility of extending his arguments in a theistic direction."[43] McDowell claims that such a broadening would lead to a worrisome ontological expansion. Ellis, on the contrary, maintains that naturalism has much to gain from a constructive engagement with theism. Thus Ellis is not content with just pointing out the deficiencies of naturalism but seeks to remedy these problems by fusing naturalism with theism. A theistic understanding of McDowell's naturalism suggests that God has granted human beings the natural capacity of recognizing values in nature, and through this recognition we can relate to and be inwardly transformed by God.[44]

Ellis's general claim is that we relate to God by relating to value, and she seeks to defend this idea from two objections: first, that we are reducing God to values and thereby transforming theism into atheism (an objection raised by some religious thinkers), and second, that there is no justification to invoke theism with regard to values (an objection raised by secular expansive naturalists).

In order to show why expansive naturalism should go in the direction of theism, and how we relate to God by relating to the domain of values, Ellis turns to Levinas. For Levinas, we relate to God by relating to value, and it is in the face-to-face moral encounter and in the "gaze of the other" that values reveal themselves to us. We are then called on to be responsible for the other.[45] The worry is that God now becomes reducible to our moral relation to others. Levinas, according to Ellis, suggested that interhuman relation is irreducibly God-involving, in the sense that it relates the subject to God. Thus to know God is to know what is to be done morally.

Ellis, however, suggests that we should not stop with Levinas's theistic naturalism, but expand this naturalism in a Christian direction. Of course, Levinas was very skeptical of the God of Christianity, because the "God who reveals Himself in the Christian manner is just one more omnipotent controller of things who wields His power with a capricious system of rewards and punishments."[46] Thus the Christian God makes us morally impotent, as everything is governed by a coercive controller. Ellis's desire to move in the direction of Christianity is motivated by the inadequacy of Levinas's theistic minimalism. We need to say something more about God in order to avoid atheism. Christianity, argues Ellis, does not present us with a God who is an omnipotent controller. Indeed, the God of Christianity is not a manipulative being but a God of love. God's divine power should be acknowledged, but we should also deny that this power is coercive.[47] God's otherness precludes the view of God as a being among other beings.[48] Hence, God is the source of nature and should therefore not "be treated as just one more item to be added to the inventory" of what exists.[49] In this way, by ruling out God as a competing cause within the causal nexus in the universe, it is also possible to reject the notion of God as a coercive controller. Ellis suggests that Levinas's critique of Christian theism is severely misplaced.

The Christian notion of God plays an important motivating role. Ellis maintains that Jesus functions as a representative and moral example for humanity rather than as a sacrificial substitute. Thus the function of the Christian conception of God is to show morally what is required of us to relate to God. Jesus is "a model of 'perfect humanity.'"[50] Levinas offers an interesting and fruitful way of connecting God with values, but his dismissal of ontological theism can be challenged, according to Ellis, by this view of Jesus.

Ellis goes on to argue that this form of expansive naturalism "grants us the right to talk about a *divinely* enchanted world."[51] Yet there is the question of how far it is possible to go in a theistic direction, and how much theism expansive naturalism can tolerate. The refusal by naturalists to fuse their ontology with theism is because God is imagined as a "separable something else which stands opposed to nature."[52] Ellis challenges this assumption by suggesting that God is both immanent and transcendent. She writes, "The difference with God, however, is that He remains radically distinct from anything within the world even whilst retaining the most intimate connection with it."⁵³ God is not reducible to the world and values, and the values that we encounter "have their source in God."⁵⁴

Ellis further argues that a theistic position is compatible with science, and that science can neither demonstrate nor exclude the existence of God, as God lies beyond the scope of scientific investigation.⁵⁵ Hence, if the expansive naturalist denies the existence of God on the basis of an alleged conflict between God's existence and science, he/she is succumbing to the same philosophical mistake as the scientific naturalist who denies the existence of values.

What difference does God make, then, for expansive naturalism? In one sense, God makes no difference at all, according to Ellis. But in another and more significant way, God changes everything. God "does so because He is the source of things."⁵⁶ God enriches nature and provides the ground for values, and God's activity is revealed at the level of values and morality.⁵⁷ Ellis is careful in not reducing God to values but maintains that "talk of values is *part* of what it means to talk about God."⁵⁸

An Evaluation of Ellis's Combination of Naturalism and Theism

Ellis offers an interesting framework for fusing a more liberal version of naturalism with robust theism. We saw earlier that panentheists fail to uphold their desired naturalism, and I concluded therefore that panentheism, be it emergent panentheism or process panentheism, is not naturalistic and cannot be considered an adequate alternative framework for moving religious naturalism forward. I am somewhat sympathetic to Ellis's approach but suggest that it suffers from some philosophical ambiguities. First, I will comment on the ambiguities in McDowell's naturalism. Second, I will argue that Ellis's application of said naturalism to values contains serious ambiguities. Third, I will comment on Ellis's theistic proposal and its alleged compatibility with naturalism.

The point of McDowell's naturalism is to show how "knowing and thought" can be considered natural phenomena, and that naturalism does not automatically entail *scientific* naturalism, with its ontological rejection of mind and values. The reason for the naturalistic rejection of mental phenomena is because of the contrast between nature, as described by science, and, "the space of reasons."⁵⁹ By "space of reasons" McDowell refers to the possibility of justifying what one says, which is an essential feature of an adequate theory of rationality. In emphasizing "space of causes" over "space of reasons," naturalism has turned a core feature of human rationality into a metaphysical anomaly. For McDowell, spaces of reasons are not only compatible with spaces of causes, but reasons can be seen as causes as well. However, this tension and this apparent anomaly is a primary source for much of the "metaphysical anxiety" in contemporary philosophical debates.[60]

McDowell suggests that "to avoid conceiving thinking and knowing as supernatural, we should stress that they are aspects of our lives" and that "they are part of our way of being animals."[61] Yet he claims that mental phenomena are not strictly physical. McDowell insists that such mental phenomena are sui generis with regard to the rest of nature, and they go beyond natural law and our current conception of the natural.[62] The space of reasons is sui generis and cannot be captured within the realm of law. McDowell, in seeking to preserve the realness and causal efficacy of the mental, seems to suggest that the causality of the mental is ontologically different from causality at the physical level. The mind, on McDowell's view, seems to occupy its own level of reality that cannot be considered *purely* physical. I am not claiming that McDowell necessarily succumbs to dualism, but he leaves us with a view of the mental that many naturalists would find off-putting. Naturalists generally seek to harmonize the level of the mental or higher-level properties in general with the level of the physical, but this attempt is resisted by McDowell, who seems to think that such harmonization invites reductionism. That there is a metaphysical grounding problem regarding the mental seems to be affirmed by McDowell. However, McDowell does not reconcile the mental and the physical but leaves sui generis mental phenomena ontologically unexplained.

It seems as if McDowell's proposal is vulnerable to "the problem of competing ontologies." In a similar way to liberal naturalists and agnostic naturalists, whom I discussed in the previous chapter, McDowell affirms the reality of the mental but does not provide a naturalistic solution for explaining how mental phenomena can be squared with the physical structure of reality. For McDowell, then, we have those objects/entities/phenomena that we can naturalistically explicate, and those that resist such naturalistic explication. This philosophical quietism concerning the mental makes it difficult to see why we should adopt a naturalistic view of the mental instead of some other nonnaturalistic framework. McDowell's naturalism is thus ambiguous with regard to the nature of mental phenomena.

The second ambiguity regards Ellis's use of McDowell in accounting for the reality of values. Ellis employs McDowell's naturalism in order to explain how human creatures interact with values, arguing that "morality is humanized in the sense that the domain it involves is essentially within reach of human beings. Our capacity to respond to this domain is a rational capacity which stems from our human nature rather than from any supernatural addition."[63] Moral engagement requires nothing more than "ethical upbringing."[64] Hence "values are part of the natural world."[65]

It seems as if Ellis confuses ontology with epistemology. That is, in outlining how humans can epistemologically know certain values to be true, she does not explain ontologically how it is that sui generis values can be a part of the natural order, or how they fit a naturalistic description of nature. She offers no naturalistic account of how to make sense of values in a purely natural world, or how it is that these values that we are brought up to believe in are not simply figments of our imagination, or constructs of our society. Ellis offers an interesting response to such complaints by saying that asking "where value fits in is misguided."[66] However, in numerous places, she states that values can causally affect us. But when doing so, Ellis is forced to give an account of how this is actually possible, and how this does not invite dualism. To merely leave this metaphysical and conceptual issue unanswered seems philosophically problematic.

Some Comments on Ellis's Theism

We shall now turn to Ellis's theistic proposal. What does she mean by "God," and how does God relate to the natural order? She makes it clear that her proposed theism goes beyond ontotheology. God is not just one more item within the "larger household of reality."[67] God, by virtue of being the creator and sustainer of everything, is not a part of the natural world, and God cannot be identified with anything in nature.[68] Ellis stresses the otherness of God, saying, "He is neither within the world nor beyond it."[69] How is God's activity imagined within Ellis's theological proposal? Ellis claims that God does not *interfere* in the natural cause of things, as God is not a competing cause among other causes. Yet she seeks to defend a robust form of theism and avoid both deism and Levinas's reduction of God to values. Hence, God must be allowed to make some difference in reality. Quite similar to the problems of panentheism, Ellis seems unsuccessful in spelling out an account of God's active presence in physical reality that could be compatible with naturalism.

Ellis is aware that the appeal to divine action is rejected by naturalists. A naturalist would presumably say that it is unnecessary to invoke God-talk "because we can explain things without this theistic detour."[70] Ellis argues that this "either/or framework"—the idea that we must either believe in divine action *or* natural causation—must be abandoned. The idea of God's omnipresence, for Ellis, means that God *does* and *does not* make a difference in reality. Because of God's omnipresence, God cannot be considered as a cause among other causes, and so, in one sense, God makes no difference at all. However, God's omnipresence also means that God is "wholly present to everything in the natural world."[71] Thus, in another sense, God makes all the difference. How do we make sense of this position? I admire Ellis's ambition to harmonize divine action and natural causation, but she seems unable to explain how this turn from *either/or* to *both* should be understood. Indeed, Ellis seems aware of the weakness of her explanation. Ellis writes concerning the problems of explicating God's active role in nature that "our inquiries must terminate inevitably and appropriately at the level of mystery."[72]

I think that this answer is unsatisfying for the concerned naturalist. A naturalist, including an expansive naturalist like John McDowell, seeks to develop a unified view of nature, whereby everything is connected to the natural domain. Everything is natural, including values. Ellis's theistic explanation of values would be considered problematic according to a naturalistic framework, as it now locates values in an extranatural realm. The reason for the existence of values is because God instantiated those values, and so we have some properties that transcend the physical domain, meaning that they do not exist by virtue of the potentialities of nature. Hence, Ellis's proposal seems to suffer from some of the same ambiguities that emergent panentheism and process panentheism struggle with. This is not to claim that Ellis's theism is untenable in itself. I am merely suggesting that there is a tension between robust theism and naturalism, which form the two halves of her proposal, that is not dealt with adequately by Ellis.

Conclusions

This chapter has evaluated three attempts to combine Christian theism with naturalism. First, I examined two forms of panentheism, and suggested that both emergent panentheism (proposed by Arthur Peacocke and Philip Clayton) and process panentheism (proposed by David Ray Griffin and Mark Johnston) are unable to avoid dualism. It was argued that emergence theory, with its emphasis on downward causation, breaks the monistic commitment of naturalism (that is, causal closure of the physical domain). I also argued that process panentheism invites dualism in proposing "mental revelation" as God's way of interacting with humanity. Because of the failure of upholding naturalism, I suggested that panentheism is not a suitable alternative framework for a religious naturalist.

Second, I examined Fiona Ellis's attempt to fuse theism with John McDowell's naturalism. I argued that McDowell's naturalism, in a similar way to liberal naturalism and agnostic naturalism, encounters "the problem of competing ontologies" by postulating irreducible, sui generis, and unexplainable mental phenomena. I further called into question Ellis's attempt to ground values in nature and argued that Ellis seems to confuse an epistemological explanation for how we come to know values with an ontological explanation of the nature of values. Lastly, I questioned the compatibility between Ellis's theism and naturalism, and suggested that the concerned naturalist would find Ellis's proposal problematic due to its possible invitation to supernaturalism.

In the next chapter we will discuss the final, and most promising, framework available for religious naturalism: panpsychism.

8

Alternative Ontology 3

Panpsychism

We have almost reached the end of our exploration of religious naturalism. Chapter 6, "Alternative Ontology 1," considered three less restrictive versions of naturalism: Mario De Caro's, Alberto Voltolini's, and David Macarthur's liberal naturalism; Colin McGinn's agnostic naturalism; and the pragmatic approach developed by Huw Price. These thinkers should be commended for admitting and exploring some of the problems created by standard, or scientific, naturalism. However, I have pointed out that both the liberal and agnostic versions are vulnerable to the problem of competing ontologies. Both perspectives consider reductionism to be untenable, and they maintain the realness of higher-level properties, such as consciousness, moral language, and meaning. Yet it remains unclear why we should take the irreducibility of such properties to specifically support a naturalistic framework.

Price's pragmatic naturalism falls short as well, but for different reasons. I argued that Price's reduction of philosophical problems to the linguistic habits of ordinary speakers fails to appreciate the seriousness of some metaphysical and philosophical issues. Additionally, Price's linguistic reductionism undermines the authority of science, and therefore his own subject naturalism.

In chapter 7, "Alternative Ontology 2," I evaluated theistic frameworks. I considered panentheism, the view according to which the world is "in" God but God is more than the world. This form of theism is, by its proponents, presented as being deeply antidualistic in the way that it seeks to overcome an interventionist conception of God, a God that when interacting with the natural order has to break the natural laws governing it. The interventionist and dualistic God is seen as scientifically, philosophically, and theologically problematic. Panentheism is considered by prominent thinkers such as Philip Clayton, Arthur Peacocke, and David Ray Griffin to be compatible with a naturalistic commitment to the causal closure of the physical. I suggested that religious naturalists, due to the metaphysical problems of monistic and pluralistic naturalism, and the failure of alternative ontology 1, might want to consider the popular position of panentheism. However, panentheism, when scrutinized more closely, seems to imply dualism, both global dualism (the ontological difference between God and the universe) and local dualism (in that it implies an ontological difference between God's actions and natural occurrences).

I explored also Fiona Ellis's attempt to combine theism with John McDowell's naturalism. I suggested that McDowell's naturalism, by leaving the human mind unexplained, encounters the problem of competing ontologies. It was also argued that it is not obvious how this form of theism proposed by Ellis fits a naturalistic framework.

In this chapter I will consider the third and final alternative ontology, namely panpsychism. I will argue that a panpsychist perspective provides a promising framework for the ongoing development of religious naturalism.

Panpsychism Today

In broad terms, panpsychism is the thesis that mind is widespread and a feature of all things. While it has enjoyed a strong presence throughout history and been favored by a number of influential thinkers, it has also in recent years been met by skepticism. Notable philosophers have described panpsychism as a view with no anchor in reality: the biological naturalist John Searle claims that panpsychism is an "absurd view";[1] the agnostic naturalist Colin McGinn argues that panpsychism is a "complete myth";[2] and the emergent naturalist Philip Clayton has criticized panpsychism for "making a robustly metaphysical move" that "cuts it off from the evidential considerations that science could otherwise provide."[3] However, it is not only naturalists who have found panpsychism to be an untenable position. The substance dualist J. P. Moreland has suggested that panpsychism amounts to nothing more than a label; it does not come close to being an *explanation* for the existence of consciousness.[4]

Despite these negative remarks, there has been an increase in interest in the panpsychist and panexperientialist frameworks for dealing with the hard problem of consciousness in particular, and higher-level properties more generally. Galen Strawson's much-discussed article "Realistic Monism: Why Physicalism Entails Panpsychism" deserves special mention due to its high impact in this debate and powerful way of framing the promise of panpsychism in light of mental realism and realistic physicalism.[5] There are, as Strawson demonstrates, good reasons for taking seriously the panpsychist notion of experience and mentality as fundamental constituents of the physical.

The "naturalistic dualist" David Chalmers has argued that we need to posit fundamental laws in order to explain consciousness and the phenomenological experience that comes with it. Thus Chalmers writes, "experience is fundamental, it does not qualify as a *physical* property."[6] Similarly, Gregg Rosenberg has argued for developing a theory of mind in relation to *panexperientialism*. This view, argues Rosenberg, is not only logically possible, it is probable.[7]

Here one should also mention Thomas Nagel's controversial book *Mind and Cosmos: Why the Materialist Neo-Darwinian Conception of Nature Is Almost Certainly False*, in which he argues against naturalistic explanations of consciousness, either via strict psychophysical reductionism or emergence theory (both being committed to a form of causal explanation). Nagel opts instead for a "postmaterialist theory" of consciousness, suggesting that evolution, if it indeed brought about consciousness, cannot be "just a physical process."[8] The explanation of consciousness, as Nagel puts it, "may have to be something more than physical all the way down."[9] This takes us in a panpsychist direction.

From a more scientific viewpoint, the neuroscientists Giulio Tononi and Christof Koch have proposed understandings of consciousness that contain panpsychist elements. Tononi, in 2004, introduced the idea of the integrated information theory of consciousness, that is, that consciousness corresponds to the capacity of a system to integrate information. Consciousness is in this view a fundamental quality and present in infants, animals, other biological systems, and even in purely physical systems, albeit to different degrees.[10] Thus, while one should not equate integrated information theory with panpsychism, it does align itself with the "panpsychist intuition that consciousness may be present across the animal kingdom, and even beyond, but in varying degrees."[11]

Despite much skepticism, it is clear that panpsychism is undergoing something of a revival within contemporary philosophy, and is also gaining attention within parts of the scientific community.

Three Arguments for Panpsychism

Several contemporary philosophers have turned toward panpsychism for explaining mentality and experience. As I argued in chapter 3, pluralistic naturalism is certainly more promising than Charley Hardwick's and Willem Drees's reductive monism. Nevertheless, an emergentist ontology does not adequately explain the emergence of consciousness. Emergence theory fails to secure the causal effectiveness of mental states and entails a problematic mysterianism. Mentality somehow emerged, but for a pluralistic naturalist this remains an unsolvable and awkward mystery.

The intrinsic problems of an emergent ontology, and strong emergence in particular, are often used as positive reasons for entertaining panpsychism. This argument comes in many forms and is often referred to as "the argument from nonemergence," "the origination argument," or "the generation problem."[12] The problem for emergentists, and similar naturalistic accommodation strategies, is to explain "why and how experience is generated by certain particular configurations of physical stuff."[13] Neuroscience and cognitive science have indeed shed light on the functions of the brain but have provided little input with regard to the issue of subjective experiencing, why physical creatures such as we are possess a first-person perspective. For some philosophers, such as Nagel and Strawson, panpsychism provides a way beyond the problematic consequences entailed by an emergentist conception of consciousness. For Nagel, the combination of mental realism, nonreductionism, and the denial of radical emergentism seem to confirm panpsychism as the best available explanation for the origin of mentality.[14] If we are realists about consciousness, if we think that mental states cannot be reduced to physical categories, nor that mental states can be construed as emergent properties, then mentality must be present at the lower levels of reality. Hence, panpsychism emerges as the winner.

Strawson reasons in a similar way. What Strawson calls "real physicalism" grants that experiential phenomena are "just physical." Yet, contrary to crude eliminativism, they cannot be fully described by physics and neurophysiology.[15] By differentiating (real) physicalism from eliminativism, it looks as if panpsychism is a live option for a physicalist who takes experiential phenomena to be "just physical."[16] For Strawson, a panpsychist reformulation of physicalism can provide new resources for addressing the origination problem of conscious phenomena.

Standard physicalism, as commonly conceived, claims the following: "[NE] physical stuff is, in itself, in its fundamental nature, something wholly and utterly non-experiential."[17] This is then combined with a second claim about experience: "[PE] experience is a real concrete phenomenon and every real concrete phenomenon is physical."[18] Strawson goes on to argue that these two claims, NE and PE, when scrutinized more closely, do not mesh, and that, on the contrary, they cause devastating tensions for physicalism.

Emergence is a useful concept when applied to relatively simple physical phenomena—liquidity is commonly invoked in these discussions. Yet the emergentist model loses its strength if we wish to understand how we get experience from something ultimately nonexperiential. Liquidity does not baffle anyone: "You can get liquidity from non-liquid molecules as easily as you can get a cricket team from eleven things that are not cricket teams."[19] The situation is significantly different when we ponder how we get subjectivity from physical interactions devoid of subjectivity. In this way emergence theory unveils the heart of the origination problem and drives many serious thinkers in a panpsychist direction.

The second argument takes its cue from the epistemic difficulties of grounding consciousness within an evolutionary narrative. It is, as we will see, closely connected to the origination argument. A version of the "argument from continuity" was formulated by William James in his seminal *The Principles of Psychology*: "*If evolution is to work smoothly, consciousness in some shape must have been present at the very origin of things.* Accordingly, we find that the most clear-sighted evolutionary philosophers are beginning to posit it there. Each atom of the nebula, they suppose, must have had an aboriginal atom of consciousness linked with it; and, just as the material atoms have formed bodies and brains by massing themselves together, so the mental atoms, by an analogous process of aggregation, have fused into those larger consciousnesses which we know in ourselves and suppose to exist in our fellow-animal."[20] James was convinced that such a doctrine of continuity "is an indispensable part of a thorough-going philosophy of evolution," whereby consciousness reaches down to the fundamental strata of reality.[21] Panpsychism corresponds well, then, to the underlying ontological theme of continuity within evolutionary theory.

Nagel takes the argument from continuity further in his most recent investigation of panpsychism. On a materialist understanding of evolution, consciousness remains an inexplicable anomaly. If materialism is true, then mental language must be expelled to a scientifically replaceable domain of "folk psychology." For Nagel, the consequences of materialism are therefore far-reaching: not only would we have to disregard our mental life as a form of illusion, we would also have to deny the reality of meaning and values. The almost-consensus in this debate is that emergence theory is the only serious option to materialism. However, strong emergence is rejected by Nagel because of its lack of explanatory power. Nagel critically notes that the emergentist scenario, according to which higher-level phenomena are instantiated by purely physical elements, "seems like magic."[22]

Due to the problems of the emergentist picture, Nagel argues that if we want to make sense of consciousness, meaning, and values, all physical constituents need to contain nonphysical components. Taking monism in a nonmaterialist direction, Nagel states that "everything, living or not, is constituted from elements having a nature that is both physical and nonphysical—that is, capable of combining into wholes. So this reductive account can also be described as a form of panpsychism: all the elements of the physical world are also mental."[23] Every physical element contains some mental aspect or protomentality, with the intrinsic capacity to combine into mental wholes and full-blown consciousness. As Nagel concludes, if evolution brought about consciousness, it cannot be deemed a purely physical process. Panpsychism alone provides the necessary ontological continuity to make sense of consciousness in an evolutionary world.

The "analogical argument," one of the most influential arguments for panpsychism, takes a slightly different form. Basically, the proponent for this way of reasoning analogically extends "certain features of our own experience to other natural forms that are judged to be similar to us in certain relevant respects."[24] Panpsychists tend to differ on what these "relevant respects" are, but certain features are often mentioned, such as the ability of S to persist through time, and the capacity of S to maintain itself within a specific environment and against certain threats.[25] This method of analogical extension was heavily utilized within the process philosophical frameworks of Alfred North Whitehead and Charles Hartshorne.[26]

Turning to today's context, we can see the presence of the analogical argument in David Chalmers's view of consciousness. Chalmers, a stern critic of reductionist strategies, opposes the idea of explaining consciousness through neurobiological explanations, or through some specific physical mechanism of the brain. No physical account, as Chalmers argues, will be sufficient for explaining conscious experiences, "as it is conceptually coherent that any given process could exist without experience."[27] This criticism applies to any physical account, for such an account fails to address the heart of the matter: "Why should this process give rise to experience?"[28] It might be the case that experience, in some way, arises from the physical. But experience is not entailed by any sort of physical configuration.

If there is an explanation for consciousness, it must be *nonreductive*. In a panpsychist manner, Chalmers suggests that we should take experience to be *fundamental* and nonderivative from other more basic stuff. Experience, in a similar way as mass and space-time for the physicist, is a fundamental property of the physical.[29] Given that the absence of consciousness is compatible with any physical theory, we need to treat conscious experience as fundamental. As mentioned earlier, Chalmers's theory of consciousness qualifies as a version of dualism, as it postulates basic experiential properties above and beyond those basic properties that belong to the domain of physics. This is a *naturalistic* dualism, however, as this position grants that "ultimately the universe comes down to a network of basic entities obeying simple laws, and allowing that there may ultimately be a theory of consciousness cast in terms of such laws."[30]

This position requires basic principles to explain how it is that experience depends on physical phenomena. As Chalmers explains, we need these basic principles so that we can "build an explanatory bridge" between the physical features of this world and experiential phenomena.[31] For Chalmers, *information* provides the missing ingredient for explaining the physical-experiential relationship. According to this double-aspect hypothesis, information has two basic aspects, a physical aspect that is publicly observable and detectable, and a phenomenal aspect that we are subjectively aware of.[32]

Thus Chalmers infers from this that "where there is simple information processing, there is simple experience, and where there is complex information processing, there is complex experience."[33] Given that information is ubiquitous, we can analogically extend experience to very simple physical structures, and thus a type of panpsychism begins to take shape. We can see, then, that Chalmers, through a double-aspect view of information processing, employs a form of analogical reasoning in order to free consciousness from the restrictive logic of reductionism, and to place conscious experience within a panpsychist and naturalistically dualist view of reality.

Between Strong and Weak Panpsychism

Panpsychism is often used as an umbrella term for emphasizing the fundamentality of mind. Hence, it is often construed as the antithesis of emergence theory, which states that mind is a secondary property in the sense that it is derivative of something more fundamental, such as matter.

Panpsychism comes in a stronger and weaker form. Strong panpsychism suggests that all matter and physical phenomena possess the kind of conscious states found in higher-level organisms. This stronger version, however, has the problem of explaining the following: "If all matter has full-blown thoughts and feelings, why do organisms need nervous systems to think and feel?"[34] It would seem as if strong panpsychism makes nervous systems, and different cognitive mechanisms, ontologically redundant. Moreover, if we take this stronger version seriously, then we are inevitably forced to ascribe conscious states to rocks, tables, cars, and so on. This is most likely one of the reasons why, as David Skrbina explains, panpsychists are reluctant to employ the stronger notion of "consciousness" in their explanatory endeavors.[35]

The weaker version suggests more cautiously that matter or physical phenomena possess mindlike properties, such as qualia or experience. This version is therefore not claiming full-blown consciousness for matter but rather a subjective aspect for the physical. It is this weaker version of panpsychism that is gaining popularity.

This weaker version comes close to David Ray Griffin's panexperientialism, which means "that all things have experience."[36] Griffin's view suggests that experience is widespread, and that mentality (such as cognition) is not an essential part of it.[37] Hence, *consciousness* is a higher-level phenomenon that arises out of lower ones, and in relation to the organizational complexity of that particular individual. Griffin employs the notion of *compound individuals* for describing an experiential unity. This kind of unity means that a multitude of experiences are interrelated into a higher-level experience, "which gives the society as a whole an experiential unity."[38] Griffin writes, "The most obvious example [of a compound individual] is a human being or any other animal with a central nervous system, in which the bodily cells are organized so as to give rise to a temporally ordered society of higher-level occasions of experience, which we call the 'mind' or 'soul.'"[39] Consequently, when there is a dominant mode of experience within a society of experiences, there is an individual. This is why a rock or a table cannot be considered an individual. On the contrary, these are "aggregational societies" in which there is no dominant member, no dominant mode of experience, and no experiential unity.[40] There is therefore a significant difference between a rock and a human being: "Whereas bodily behavior of a human being is directed by purposes of a series of dominant occasions of experience, the behavior of a rock is simply a statistical result of the average effect of its billions of members."[41] A rock is a mere aggregational society with no experiential unity and so no ontological individuality or subjectivity.

It will become clear that a weak form of panpsychism is compatible with weak emergence. Weak panpsychism still allows for the emergence of higher-level forms of experience, such as consciousness. In this way, weak panpsychism can agree with emergence theory that *robust* consciousness is a higher-level phenomenon. However, weak panpsychism rejects the strong emergentist notion of *novel phenomena*; that is, phenomena that arise unexpectedly, are irreducible with regard to the physical base level, and exhibit certain properties, such as causal powers, that cannot be identified at the lower levels (*PON1–5*). Novel phenomena, as we have seen, signify ontological newness. For the weak panpsychist, consciousness is not ontologically novel but is present at the lower level, albeit to a lesser degree.

Griffin's formulation of panexperientialism is helpful for outlining the fundamentality of the mental. I am largely sympathetic to Griffin's approach. Yet I am reluctant to label my own approach "panexperientialism," as I think that Griffin's theory is too focused on experience. I suggest that the subjective dimension that Griffin seeks to retain and protect from reductionism might include a variety of phenomenal properties, or aspects of qualia (subjective experiences). For this reason I will still employ the label "panpsychism" throughout this chapter. Panpsychism, as I use it, emphasizes the fundamentality of mindlike properties, including experience, other phenomenal properties, and aspects of qualia. In this way, I am proposing weak panpsychism, rather than strong panpsychism, and the possibility of combining weak panpsychism with emergence theory in its weaker form.

Below I will investigate the panpsychist dimension of emergence theory, as construed by prominent religious naturalists.

The Panpsychist Dimension of Emergence Theory

In this section I will develop the idea that emergence theory, as it is expressed by religious naturalists, seems to invite a panpsychist understanding of consciousness. It will become clear that Stuart Kauffman and Loyal Rue assume a metaphysical view of emergent phenomena/properties that brings emergence theory close to panpsychism. While I was editing this chapter, Kauffman published an article and a book in which he explicitly endorsed a form of panpsychism. I cannot claim that I could foresee Kauffman's advance toward panpsychism, but, as I will show, his overall stance on emergence theory and biosemiotics invites a panpsychist understanding of reality. I will examine this panpsychist move in light of Kauffman's emergence theory. We can also see panpsychist tendencies expressed by both Ursula Goodenough and Donald Crosby in relation to their commitments to biosemiotics.

Emergence theory and panpsychism have long been thought of as two diametrically opposing perspectives; it is my hope to change this view. It will become apparent that there is a strong connection between emergence theory and the (relatively speaking) new science of biosemiotics. I will argue that biosemiotics, with its emphasis on the mindedness of nature, increases radically the plausibility of a panpsychist ontology.

Agency at the Fundamental Level: Considering Kauffman's View of Molecular Autonomous Agents, and Rue's Inherentism

As a part of his articulation of emergence theory, Kauffman ascribes agency to molecules and bacteria in order to show how agency is a fully natural capacity. I will outline Kauffman's view of agency more thoroughly, which will make it easier to spot the similarities between emergence theory and panpsychism. It will also make it clearer and less surprising why Kauffman has moved in the direction of panpsychism.

Stuart Kauffman's goal is to identify "the minimum natural system to which it makes sense to attribute teleological explanations" and the language of agency.[42] Kauffman proposes a five-part criterion for a minimal molecular agent: it should be able to reproduce; it must perform at least one work cycle; it must have boundaries, and it must be possible to individuate the agent; it must engage in self-propagating work; and it must be able to *choose* between at least two alternatives.[43] "Choose," according to Kauffman, is a teleological term, and in this way it is possible to say that bacteria and molecular agents exhibit goal-directed behavior, similar to how human agents choose and act. By identifying the minimum requirements for agency, we can then by extrapolation explain how we get something like human consciousness: "We take agency to be a matter of *degree*: some minimum conditions must be met for one to ascribe it at all, and then it increases (roughly, as a function of the increase in organizational complexity through evolution) until one encounters full, robust conscious agency."[44] Kauffman has recently started to move in the direction of semiotics and is now convinced by the semiotician Kalevi Kull that biosemiosis is real in the universe.[45] The concepts of "function," "doing," and "purpose" are quite common in biology. However, Jacques Monod claimed that there are only happenings, no "doings," within biology and physics. An organism's behavior is a mere "as if teleology," which Monod called "teleonomy." We humans, according to Monod, perceive "doings" and purposes in nature, but this is merely an illusion brought about by natural laws: "There is only the mechanical, selected appearance of 'doing.' "[46]

Pointing toward the bacterium swimming up the glucose gradient to gather food, Kauffman maintains that teleology and "doing" is as real as it gets. Thus he claims that " 'doings' are real in the universe, not merely Monod's teleonomy."[47] He refers to some of Kull's work that shows the presence of *semiotic codes* in cells, by which "the cell navigates its 'known' world, 'known'—without positing consciousness—via the code and, in general, probably evolved by selection, encoding of the world as 'seen' by the organism."[48]

Likewise, Rue ties mentality and agency to the levels of the organizational structure of living systems, which are "genuine teleological phenomena *within* the universe."[49] There is, in Rue's view, "a universal telos embodied in all living things," and "agency and motivation can be claimed as universal characteristics of life."[50] In this way, "all living things have it in their nature to seek the ultimate goal of viability, each according to its inherent strategies."[51] Rue further argues, drawing on the research by the emergence theorist Terrence Deacon, that even nonliving systems reveal teleology and agency, for example, an autocell, which is a theoretical construct of an emergent property.[52] For Rue, similar to Kauffman, the activities of a bacterium make it possible to talk about agency and teleology within nature but without employing dualistic categories or concluding that this teleology must have a transcendent source.

What is interesting about the emergence theory proposed by Kauffman and Rue is that they seek to explain how we get something like consciousness and values later in evolution by showing how higher-level phenomena are already present within the fundamental constituents of nature. That is, in order to account for the physical-mental transition, these emergence theorists seek to show how mentality is already present within the physical constituents that make up mind, and so allow for agency. Hence, Rue and Kauffman take a different approach compared to those who might be referred to as *combinatorial emergentists*.[53] Someone who adopted a combinatorial view of emergence would maintain that agency, consciousness, and human causality are real and novel emergent phenomena that arise due to the combined effects of the physical constituents, constituents that themselves possess no mentality or agency. Galen Strawson explains nicely the emergentist position: "Physical stuff in itself, in its basic nature, is indeed a wholly non-conscious, non-experiential phenomenon. Nevertheless, when parts of it combine in certain ways, experiential phenomena 'emerge.'"[54] According to emergence theory traditionally construed, consciousness, mind, or awareness are not intrinsic properties; they are resultant properties. We can see that Kauffman and Rue depart from the traditional emergentist position by postulating agency at a more basic level in order to more fully explain higher-level forms of agency.

Kauffman's and Rue's way of accounting for agency, and other higher-level phenomena in nature, is quite similar to how David Ray Griffin approaches the body-mind debate. On the basis of his panexperientialism, Griffin argues that science reveals that experience is not limited to higher-level animals. We find experience in bees, bacteria, and single-celled organisms such as amoebae and paramecia. Today DNA molecules are viewed more as organisms than machines, and so Griffin suggests that even macromolecules might have some kind of experience. Moreover, Griffin, referring to the work of the philosopher William Seager and the physicists David Bohm and Basil Hiley, maintains that quantum physics offers us good reasons for attributing mentality to the elementary units of nature.[55]

Griffin concludes from this, in a similar way to Kauffman and Rue, that mentality comes in *degrees* and manifests differently in different organisms, depending on the level of complexity. Griffin writes, "However, this emergence is (by hypothesis) only different in degree, not different in kind, from the emergence of living occasions in eukaryotic cells out of macromolecules and organelles."[56]

Given Kauffman's view on molecular agency, it might not be too surprising that he is now affirming panpsychism. In a new article titled "Cosmic Mind?" Kauffman explores the possibility of panpsychism in light of *the measurement event* in quantum mechanics. This event reveals the "wave/particle duality of quantum mechanics" whereby light and matter can appear as both waves and particles. Measurement, argues Kauffman, may be "testably mediated" by human intent and consciousness, even when the human participant is distant.[57] Kauffman cautiously suggests that this might be a form of "psychokinesis," where the mind can alter the physical world.[58] However, if this relationship between the observer and the observed is correct, then one must ask the following: "But if true, measurement has happened since the Big Bang, without human mind. Thus perhaps measurement is mediated by mind at many levels—perhaps down to the idea that quantum variables can measure one another, perhaps consciously and with free will."[59] Kauffman introduces here a dualism between *res potentia*, which refers to ontologically real possibilities, and *res extensa*, which connotes ontologically real actuals. This dualism is "a world of both possibles and actuals, linked by measurement in which mind mediates measurement."[60] These possibles, located either in space or elsewhere, change instantaneously on external observation and measurement. In turn, this measurement/observation produces another actual, a change in the becoming of the universe, and so another possible.

Kauffman suggests that if measurement is ontologically indeterminate, then "measurement converts possibles to actuals."[61] This entails a new view of reality, as mind and matter now "interact in the becoming of the universe via measurement."[62] This, argues Kauffman, leads to a participatory panpsychism. From this, Kauffman extracts a *triad*, a triad that "consist[s] in actuals, possibles, and mind measuring possibles, to yield new-in-the-universe actuals."[63] Quantum mechanics boils down to this triad.

The participatory nature of the universe is further revealed in the quantum enigma, when we prepare an electron in a superposition. The remarkable thing is that we can "look in box 1 to measure if that electron is in box 1 and if yes, the electron comes to exist in box 1."[64] Moreover, if we look and measure the electron in box 1, and the electron is absent in box 1, then it ontologically appears, or comes to exist, in box 2 "by null measurement."[65] It seems as if Nature and humanity create reality together, and that humanity is an integral part of the becoming of the universe. Hence, we can here detect the dual nature of the relationship between *res potentia* and *res extensa*, where mind mediates this relationship.

Kauffman further elaborates on the possibility of cosmic minds/mind in light of the measurement event. We have a set of entangled particles, and "each alters the wave function of the remaining entangled particles instantaneously. The more particles are entangled, the more correlated are their measurement outcomes. If bursts of consciousness arise at measurement, entangled particles might have a kind of 'unity of consciousness.'"[66] It might be possible, argues Kauffman, to apply this measurement event, or observer effect, to the universe as a whole. The universe at measurement can observe and choose among the possibles relating to the outcome of a measurement. The universe mediating a measurement would give rise to new actuals, thus changing its own direction.

Even souls are possible in this new post-Cartesian and post-Newtonian view. If entangled variables (with phenomenological features) are imbued with mind, or give rise to bursts of consciousness, then "such entangled variables just might be something like souls that are richly minded, perhaps set free in space to join other clouds of cosmic mind(s)."[67] Kauffman, however, remains cautious and suggests that souls and cosmic minds are unlikely but not impossible given this quantum view of reality and the mind's causal contribution to the ontological becoming of the universe.

The Semiotic World of Goodenough

Following a semiotic approach to nature, Ursula Goodenough maintains that physical reality is infused with semiotic systems and awareness, and that it is far from mindless. Biological organisms are, in general, endowed with semiotic information. Single-celled organisms and multicelled organisms stand in a relationship with their respective environment and carry interpretable information. Numerous systems, according to Goodenough, can be said to mediate *cellular awareness*. She writes, "Many of the encoded traits make use of receptors (like the amoeba's receptors for decaying-food molecules) that detect relevant signs in the environment and convey their meaning to the organism."[68] Throughout nature, from single-celled to multicelled organisms, we see detection of relevant signs and appropriate reactions. Mind is not an anomaly in this view; it is a fundamental feature that is present in the most basic biological entities.

Moreover, all creatures, and "even the first proto-life, are endowed with sentience systems that pay attention to relevant parameters."[69] Sentience, in the semiotic model proposed by Goodenough, is defined in terms of receptor-based awareness, signal-transduction, and appropriate responses. This form of sentience is evident in proto-life, single-celled, and multicelled organisms.

Goodenough points out that experience has often been restricted to higher-level organisms. One common argument is that experience requires a developed nervous system. We humans, according to her, "carry a strong bias here because we each hold an I-self account of what it's like to experience our version of experience."[70] The problem is that we do not know what it is like to be any other organism. We lack the epistemological access to the perspective of other more simple organisms. However, when we observe the behavior of simpler creatures, such as insects, the idea that experience requires a nervous system becomes problematic.

Goodenough introduces also the concept of *nociception* to demonstrate the experiential character of many low-level organisms that lack nervous systems. This concept designates "the ability to detect and respond to aversive-noxious environmental stimuli."[71] Such systems are found in bacteria and single-celled eukaryotes, and the stimuli within such systems are mediated by receptors and signal-transduction cascades that require no nervous system.[72] In this way, Goodenough concludes that nonneural organisms can be said to have experience, as they interpret their respective environments, despite the lack of a complex nervous system. Goodenough, alongside Kauffman and Rue, posits agency/experience at a basic level of nature, and so opens up the possibility for a panpsychist ontology. This will become even clearer when I later on connect panpsychism to pansemiosis.

Crosby and the Thou of Nature

In one of Donald Crosby's more recent works he calls "attention to the inward, first-hand, and thus 'thou' character of forms of life."[73] Similar to Kauffman, Rue, and Goodenough, Crosby seeks to show how agency is present in basic life forms, which takes us beyond the modernist conception of nonhuman life as mindless. Science, based on the modernistic tools of external observation and quantification, is unable to grasp the "inwardness that is characteristic of all life, even in its most primitive forms."[74] Rather than assuming a great ontological divide between human and nonhuman life, we must come to understand that the difference between us and other life forms is one of degree, not kind or categorical distinction.[75] Nature is not mindless; it is "an interconnected system of thous reaching far beyond human thous."[76] Consequently, there must be continuity between matter and mind. However, this view is often met by skepticism. Crosby quotes zoologist Donald Griffin: "Most of Darwin's basic ideas about evolution are now generally accepted by scientists, but the notion that there has been evolutionary continuity with respect to conscious experience is still strongly resisted. Overcoming this resistance may be the final crowning chapter of the Darwinian revolution."[77] The phenomenon that is human consciousness is a product of a gradually increasing capacity for sentient awareness and mental capabilities already possessed by nonhuman life forms. Crosby, in arguing for the thou character of nature, turns to Evan Thompson. Thompson in his book *Mind in Life* seeks to bridge the gap between biological life and consciousness. He defines life, or living systems, in the following way: a living system shows *autopoiesis*, that is, the system is self-making. An autopoietic system, according to Thompson, is one "that actively produces and maintains from within itself and by means of its own internal resources" a boundary between itself and its environment.[78] Moreover, an essential feature of a living system is *cognition*, or sense-making, whereby an organism/system identifies its environment and adapts accordingly. Cognition is the "remarkable capability of life-forms to identify, adapt to, and in many cases alter their environments."[79]

From the system's ability for self-making and sense-making follow other features that are characteristic for different life forms, namely sentience (the feeling of being alive and self-awareness), intentionality (the ability of the organism to envision the world and "act with some degree of awareness in reference to it"), and purpose (the ability to act toward certain goals).[80]

Based on Thompson's observation regarding the intrinsic inwardness and qualitative firsthand experience of all life forms, Crosby concludes that mind and consciousness, being products of evolutionary development, are not restricted to human beings but are widespread in nature.

Despite Crosby's ambition to show the robust presence of mind in nature, he still rejects the idea of mind or teleology being inbuilt into the universe. For him, mind is still emergent and dependent on the potentialities of matter, and teleology is not "a primordial feature of the universe."[81] Crosby therefore rejects Nagel's panpsychism. However, given the problems facing pluralistic naturalism (as shown in chapter 3), I suggest that Crosby should seriously consider the panpsychist option in order to properly ground his teleological vision.[82] Indeed, as will be shown, those who subscribe to biosemiotics should go in the direction of *pansemiosis*, and therefore panpsychism.

Biosemiotics, Emergence, and Panpsychism

Thus far we have seen how religious naturalists such as Stuart Kauffman, Loyal Rue, Ursula Goodenough, and Donald Crosby, all proponents of emergence theory (or pluralistic naturalism), see nature as infused with agency, mind, and teleology. Furthermore, Kauffman, Crosby, and Goodenough explicitly endorse a biosemiotic approach for understanding mind and meaning in a natural world. Goodenough claims awareness and cognition for proto-life, and suggests that experience is not dependent on a nervous system. We have also seen that Kauffman in a recent output has more explicitly embraced a version of panpsychism.

In light of the interests that these religious naturalists have shown in biosemiotics, I will now briefly outline this field of research and show how the science of semiotics justifies and even gives some empirical credibility to a panpsychist worldview.

Biosemiotics

The guiding principle behind biosemiotics, roughly speaking, is that what distinguishes life from nonlife is the "instinctive capacity of all living organisms to produce and understand signs."[83] A sign is something that stands for something other than itself, and a sign is any physical form that has been made externally to stand for some object, known as a referent or referential domain.[84] For example, in human life, signs "allow people to recognize patterns in things; they act as predictive guides or plans for taking actions."[85] Semiotics is the attempt to locate signs and the representational capacity of living organisms in nature. Science has had a long tradition of merely studying the interactions between physical constituents or material substrates. Hence, according to the "traditional model," organisms do not stand for anything, and there is no reason to invoke the categories of "agency" or "subjective experiences."

Semiotics is the science of understanding the production and interpretation of signs. Thus *bio*semiotics is the attempt to fuse the general framework of semiotics with modern biology and subsequently expand the horizon of science. Biosemiotics studies the production and interpretation of signs in the biological world, between molecules, proteins, and different organisms. This approach has two main rivals: physicalism, which rejects semiosis in both the organic and physical world, and pansemiosis, which extends semiosis and accepts it even within nonliving physical systems.[86]

Although biosemiotics remains the more favored position among religious naturalists, I will give a brief argument for thinking that pansemiosis is a real possibility that opens up for panpsychism. Moreover, it shall be seen that pansemiosis is useful for framing the compatibility between weak panpsychism and weak emergence.

Restrictive Biosemiosis or Pansemiosis?

One could say that semiotic processes reach far down into nature, but not *all* the way down. Perhaps they are present only in biological life forms (biosemiosis) and not in purely physical systems (pansemiosis). Thus semiotics is of little help to a panpsychist project. Some argue, for example, that before the advent of life on Earth, there were no signs and no representational capacities. If this is true, then semiosis is not universal or ubiquitous.

The notable emergence theorist and biosemiotician Claus Emmeche has published an interesting paper that gives some credence to thinking that sign processes are present at both the biological and a (general) physical level of reality.

Emmeche seeks, through biosemiotics, to explain the origin of consciousness in evolution. The main task is to show how particular physical properties have also acquired irreducible experiential qualities, the "inner experiential sphere," or what Jakob von Uexküll simply called "Umwelt."[87] Scientists have consulted biology, complexity research, and cognitive science, but, according to Emmeche, failed in appreciating the relevance of a biosemiotic approach in unveiling the mystery of consciousness in nature. Emmeche links sign processing with experience, and because biosemiotics reveals and explains the presence of signs throughout nature, it can also provide insights into the reality of experience in nonliving physical entities. Thus semiotics may well hold the key for understanding mind in nature.

Emmeche opens up for a panpsychist understanding in two ways. First, he introduces the idea of *protosemiosis* in nonliving systems. Semiosis is not restricted to the living world but occurs in nature as a whole. For example, when oil is heated "in a frying pan one can see the formation of heat convection cells."[88] Furthermore, the "oil that is organized into these 'cells' (which are far from being alive in any biological sense) can be understood as a form of 'proto-semiotics,' or physiosemiosis (as sign activity occurs in the non-living chemical and physical realm, 'in the background' as it were, 'throughout the material realm.')"[89] Thus even in purely physical systems what "is expected to be found in lower intensities are specific sign processes, that is, signs producing and mediating other signs."[90] In this way, we can see that experience, defined as "particular significant interactions between a system and its surroundings," goes beyond biology, and so semiotic activity is more than the form of activity associated with living biological systems.

Is this panpsychism? Emmeche rejects this conclusion and argues that consciousness is something we see later in evolution; hence, consciousness is an emergent phenomenon, not a basic property. Thus this idea "is not some form of panpsychism, according to which consciousness should be found . . . in purely physical systems."[91] However, he also writes, "in a very general sense of the word *experience* one can say that all these systems, even purely physical [ones], *experience* something, get 'irritated' or affected by their surroundings, and store this influence, even when such stimuli are quite evanescent or produced by chance."[92] Experience is therefore widespread.

Contrary to Emmeche's negative conclusion, I suggest that this fits rather well with panpsychism, and in particular with the form of panpsychism under consideration. The problem is that Emmeche seems to think that panpsychism is required to postulate full-blown consciousness at the fundamental level. This is a gross but rather common misunderstanding of panpsychism. One could, and I have suggested that we should, adopt a weaker form of panpsychism, suggesting that experience and mindlike properties (such as phenomenal properties) are present at the fundamental level of reality. If we adopt a weaker form of panpsychism, the similarities between Emmeche's pansemiosis and panpsychism become noticeable. Indeed, David Chalmers's take on panpsychism bears a strong resemblance to the semiotic framework proposed by Emmeche. As we have seen, Emmeche stresses the omnipresence of experience; Chalmers does the same. Chalmers suggests that experience is ubiquitous and that it "is probably *everywhere*."[93] Moreover, they both link experience to information processing; an entity, whether complex or very simple, can interpret its environment and act accordingly. Hence, wherever there is information, there is experience, and where there is experience, we have the precursor for robust consciousness.

Emmeche's helpful exploration of biosemiotics provides an excellent resource with which religious naturalists can approach panpsychism. On this view, experience is a fundamental category of reality, present in both living and nonliving things (panpsychism), while conceding the fact that consciousness is a later development in evolution. Semiotics, which as we have seen is coupled with emergence theory, clears a path to a panpsychist understanding of reality. In this way, we can see that weak panpsychism and weak emergence theory are compatible.

Metaphysical Objections to Panpsychism

We have now seen how emergence theory, together with a biosemiotic approach and pansemiosis in particular, opens up for a panpsychist understanding of nature. Nature, on this view, is infused with mind, and the emergence of conscious creatures is not about getting "something from nothing" but "something more from something else." As we have seen, despite the resurgence of panpsychist thinking, the idea of mindlike properties as fundamental is not popular among contemporary philosophers. Indeed, some have argued that panpsychism is not only counterintuitive but faces insurmountable problems. This section is devoted to exploring some of these objections and metaphysical issues associated with panpsychism. I am not trying to show that panpsychism is a problem-free position. Not at all. I dare to say that all ontologies have their fair share of problems. My aim is more modest. I seek to show that this is a metaphysical approach that is worth taking seriously, and that it should not be so easily ruled out as a metaphysical framework for dealing with the hard problems of philosophy. Moreover, religious naturalists ought to seriously consider panpsychism as a possible option, both because of its metaphysical strengths, and because it can motivate a religious response toward nature.

In the following sections, I will consider some objections to panpsychism. I will explore the infamous combination problem for panpsychism. This refers to the puzzle of how microconsciousness can combine into macroconsciousness. Others have suggested that panpsychism is not compatible with a naturalist ontology. Lastly, it has been argued that panpsychism, at least in its weaker version, is a meaningless position.

The Combination Problem

The combination problem is by many considered to be *the* argument against panpsychism. William Seager coined the term "the combination problem" and described it as "the problem of explaining how the myriad elements of 'atomic consciousness' can be combined into a new, complex and rich consciousness such as that we possess."[94] This problem, says Seager, goes back at least to William James and his famous work *The Principles of Psychology*. Seager further argues that the "combination problem points to a distinctive generation problem in panpsychism which is formally analogous to the problem of generating consciousness out of matter."[95] Panpsychism, then, will have no advantage over physicalism if it falls prey to essentially the same metaphysical problem.[96]

Contrary to Seager, I do not consider the problem for panpsychism and the problem for physicalism pertaining to the origination problem of consciousness to be strictly analogous. The problem for physicalism is that the ontological categories associated with subjectivity (first-person perspective) seem awkwardly out of place if we take only the categories of the physical sciences to be real. That is, physicalism encounters an *ontological* problem.

If we consider instead the problem for panpsychism, it seems to be an *epistemological* problem, namely how to explain or offer an account of the process whereby micromentality combines into large-scale mentality. The argument does not state that this micro-macro transition is ontologically impossible, just that we cannot explain this in a systematic manner. Thus I agree with David Skrbina that "the Combination problem is perhaps better seen as a call for details."[97]

Philip Goff insists that the combination problem is ontological, that the difficulty of providing an explanation is due to the fact that microconsciousness cannot form macroconsciousness or an integrative whole. He argues that the emergence of "macroexperiential properties from the coming together of microexperiential properties is as brute and miraculous as the emergence of experiential properties from non-experiential properties."[98] Goff thus suggests that microexperience and macroexperience are as different from each other as mind and matter. Goff also writes, "Whatever sense we can make of experiences summing together, it is contradictory to suppose that the experiential being of lots of little experiencing things can come together to wholly constitute the *novel* experiential being of some big experiencing thing. Even the experience of a severely pained subject of experience is sufficiently different from the experience of slightly pained subjects of experience as to make it incoherent to suppose that the former could be formed from the latter."[99] Goff's conclusion is that the panpsychist explanation is incoherent because we cannot get organizational wholes from simpler parts.

It should be made clear that this argument is aimed not exclusively at panpsychism but also at any theory that maintains a holistic view of reality, that "the whole is more than the sum of its parts." Consequently, Goff's antiholistic remark leaves us with not only ontological reductionism but eliminativism (we have to deny the reality of the whole). If one must reject holism and the reality of irreducible organizational wholes in order to escape a panpsychist conclusion, then the price seems to be awfully high. Following Goff's view, higher-level properties would be nothing but the complex arrangement between physical parts. We are left with a bleak view of reality that most, if not all, naturalists that I have discussed should find troublesome. Religious naturalists should therefore not be completely put off by this objection.

An Ontological Solution to the Combination Problem?

Sam Coleman has argued that the combination problem is ontological, and that the real problem is how to explain how microsubjects can combine into macrosubjects. A central idea of panpsychism, according to Coleman, is that fundamental entities possess subjectivity and that subjects, with their unique experience, combine or give rise to a macrosubject with a unified experience. This "constitutive panpsychism," as Coleman calls it, maintains that "human experiencers are at bottom constituted by materially fundamental 'micro-subjects'—corresponding to basic physical particles—who are the first holders of the phenomenal qualities we enjoy atop our macroscopic perches."[100]

Yet this idea immediately raises an issue regarding the relationship "between the *subjectivity* of these fundamental subjects and the unified, single-perspective of the human subjects they compose."[101] We have before us, says Coleman, a *subject* combination problem. The problem is that microsubjects, however arranged, do not yield the existence of a unified macrosubject, such as human consciousness.

Rather than completely rejecting panpsychism, Coleman seeks to modify it. He suggests that we abandon the idea of subjectivity at the fundamental level, and instead come to view it as a "structural property," similar to an emergent phenomenon such as H_2O. This structural view therefore denies the fundamental nature of a subject. Coleman is, however, clear on the fact that this "panpsychism" is brought dangerously close to physicalism: naturally, "it won't be surprising if physicalists find the end position significantly less off-putting than panpsychism."[102] Coleman's end position is neutral monism, which states that "unexperienced qualities permeate basic matter. Certain portions of matter exhibit a configuration which provides awareness of the qualities they bear."[103]

I want to suggest that Coleman's position is consistent with the kind of panpsychism that I am striving for. The reason for Coleman's move away from panpsychism (and toward physicalism) is that he assumes that phenomenal qualities, like experience, have to be tied to a subject. That is, an experience requires an experiencer. Given that panpsychism is committed to a view of subjects at the fundamental level, and that subjects cannot combine, we have to opt for a nonpanpsychist solution.

Contrary to Coleman, I suggest that panpsychism, for it to still be worthy of its name, does not require the rich notion of subjects. More controversially, I want to suggest that the idea of noncognitive experience is coherent. I agree with Gregg Rosenberg, who writes, "The privacy of consciousness forces us to build a kind of tolerance for alien experiences and feelings: A manta ray sensing the electromagnetic structures on the ocean floor may experience qualities we could never imagine. We also have to allow that simpler and simpler organisms may have experiences of simpler and simpler kinds, as well as alien kinds. So the open-ended character of the concept requires us to accept that there could be experiences both very alien to, and much simpler than, any we can imagine."[104] Given the open-ended character of experience, that experience is present in very simple organisms, it seems as if experience outruns cognition, "that experience exists throughout nature and that mentality (i.e., a thing requiring cognition, functionally construed) is not essential to it."[105]

According to my offered panpsychism, subjects are therefore not fundamental properties but emergent features of reality. This is akin to what Griffin calls "compound individuals," that is, individuals "in which a higher-level experience (which might be conscious) emerges."[106] Recall also Goodenough's argument that I outlined earlier. She claimed that when we consider the presence of experience in lower-level organisms, bacteria, and even single-celled organisms, we can see that a nervous system is not a necessary component for experience. This might be seen as an argument for the idea that experience, in some sense, outruns cognition and specific physical mechanisms.

To conclude, panpsychism is not committed to subjectivity at the fundamental level but allows for it as a higher-order feature. In this way, Coleman's critique is helpful in outlining the difference between weak panpsychism and strong panpsychism, in that his argument only defeats or affects the stronger version. Weak panpsychism, which I propose should also be combined with weak emergence, escapes Coleman's critique.

Compatible with Naturalism?

Panpsychism postulates mentality, or experience, at the very fundamental level of reality and considers mental properties as on par with basic physical properties, such as mass and energy. Because of this, some critics argue that panpsychism is no longer committed to a *naturalistic* ontology, as it strays too far from what historically has been considered a naturalistic point of view. J. P. Moreland writes, "If a naturalist posits mental properties or potentialities as basic . . . he has opted for a version of panpsychism and abandoned naturalism."[107] In Moreland's view, we adopt either panpsychism *or* naturalism. Historically speaking, naturalism has been combinatorial, mechanistic, and physicalist. Hence, panpsychism is the direct antithesis of naturalism and cannot be a version of it. I find this historical argument quite unconvincing. Simply because a position has traditionally been associated with certain ideas does not necessarily mean that it cannot be developed in new directions.

One could perhaps argue that naturalism, given its very definition and association with mechanistic thinking, is something that cannot allow for panpsychist elements. However, as D. S. Clarke has noted, "There are no grounds for this identification [between mechanistic thinking and naturalism] because there is no part of the definition of 'naturalism' that precludes its extension to panpsychism."[108] A definitional exclusion of panpsychism seems unsuccessful. We need some positive reasons for thinking that panpsychism cannot be a viable option for the naturalist. Perhaps someone would argue that panpsychism betrays some core thesis of naturalism?

Owen Flanagan has explored some "varieties of naturalism." Going through different forms of naturalism, Flanagan suggests that antisupernaturalism seems to form "the common core, the common tenet, of 'naturalism.'"[109] The one thing that "all card-carrying naturalists should accept," as a necessary feature of a naturalistic worldview, is the rejection of supernaturalism.[110] Thus if panpsychism is going to be acceptable from a naturalistic perspective, it cannot entail supernaturalism. It is true that panpsychism, past and present, has been coupled with nonphysical and supernaturalistic ideas. Panpsychism, however, "does not itself require the introduction of nonmaterial elements."[111] I suggest that panpsychism is friendly to supernaturalism/theism, and that it can be extended in such a direction (as I will argue toward the end of this chapter), but panpsychism as such does not necessitate any nonphysical entities. Panpsychism can be considered an option for naturalists, including religious naturalists, as it does not logically entail supernaturalism.

It has also been suggested by Willem Drees that a panpsychist or panexperientialist outlook denies the evolved character of sentience and subjectivity, and for that reason cannot be considered naturalistic in any meaningful sense. Moreover, Drees states that panexperientialism "is substantially at odds with current science, where the disciplinary order . . . seem[s] to express insights about the layered character of reality."[112]

First of all, I fail to see why sentience *has* to evolve according to a naturalist ontology. From a naturalist view, some properties will be taken as basic properties, and my proposal is that experience (which Drees equates with sentience) can be viewed as such a property. Drees, however, also maintains that taking sentience as a basic property is "at odds with current science."[113] I confess that my proposal goes beyond science, but so do all ontologies, including Drees's own (hence, he calls it a low-level metaphysics). Nevertheless, it does not go *against* science. On the contrary, given that some areas of science are moving in the direction of viewing agency as widespread in nature (such as biosemiotics), and clearly rejecting the idea that only higher-level creatures possess sentience, a form of panpsychism is to be considered not just a possible conclusion but even probable.

Secondly, even if one assumes that sentience has evolved, I think that panpsychism is not undermined. It is true that a panpsychist denies the emergence of sentience *if* we take this emergence to be experience from nonexperience. However, the type of panpsychism that I am proposing can still claim that experience, or mentality, manifests differently depending on organizational complexity. That is, mentality does exist at the fundamental level but will be different in different organisms depending on their respective physical complexity. We can then say that mentality has evolved, at least to an extent.

I therefore agree with the panexperientialist Gregg Rosenberg that an ontology that posits mentality at lower levels of reality can be considered naturalistic. This is, of course, not a physicalist version but, as Rosenberg calls it, a "liberal naturalism." Therefore, religious naturalists can maintain their commitment to naturalism while adopting the more expansive ontology of panpsychism.

Is Weak Panpsychism a Meaningless Position?

I stated in the beginning of this chapter that I reject strong panpsychism and the idea that robust consciousness is present at the physical base level. Instead, I opt for weak panpsychism, suggesting that mindlike properties such as experience are fundamental properties. Some argue that whereas strong panpsychism is untenable, weak panpsychism is meaningless. It is claimed that weak panpsychism is too soft and too vague when it comes to specifying the relationship between the mental and the physical, and the nature of the mental. Colin McGinn has given voice to this concern. Weak panpsychism, argues McGinn, amounts to the idea that physical entities possess not full-blown consciousness but *protomental* states. Such protomental "states can *yield* conscious states while not themselves *being* conscious states."[114] Physical states are not fully mental, but when combined in particular ways they can give rise to mental states and conscious phenomena. McGinn critiques this version of panpsychism for being empty. He writes, "But the weak version merely says that matter has *some* properties or other, to be labeled 'protomental,' that account for the emergence of consciousness from brains. But of course *that* is true! It is just a way of saying that consciousness cannot arise by magic; it must have some basis in matter."[115] McGinn's skeptical conclusion is that weak panpsychism effectively collapses into plain emergence theory, whereby mind is dependent on some property of matter for its instantiation. This critique affects some versions of panpsychism. David Chalmers's notion of *protophenomenal properties*, for example, seems rather vulnerable to McGinn's critique.[116] It should be noted that the version of panpsychism that I am defending, and which comes close to panexperientialism, does not merely posit "protoexperience," or something like it, at the fundamental level; that is, the notion of protophenomenal properties that themselves are nonexperiential but that produce conscious phenomena when combined in the correct way. Instead, I suggest that basic properties themselves are phenomenal and can be considered experiential in essence. This panpsychism is significantly stronger than the one that McGinn is critiquing and so should not be dismissed as meaningless or semantically vacuous.

The Religious Relevance of Panpsychism

It is time to turn to the religious aspect of panpsychism and see if it can provide a religious vision of nature and physical reality, and so serve religious naturalists in this way. I will explore the nontheistic formulations of religious panpsychism of D. S. Clarke, Freya Mathews, and Stuart Kauffman. A theistic understanding of panpsychism will also be investigated. Such a theistic approach to panpsychism is proposed by Pierre Teilhard de Chardin and Keith Ward.

Nontheistic Religious Panpsychism: D. S. Clarke, Freya Mathews, and Stuart Kauffman

D. S. Clarke suggests that in order to understand the connection between a religious attitude and panpsychism, we must first "abstract a common religious attitude from the various forms in which it is expressed with the world's religions."[117] The religious attitude, says Clarke, is the affirmation of mentality as eternal, traditionally connected with God or other supernatural beings. In religious traditions, our affirmation of the ultimate or the eternal is of greater importance than our individual lives. Hence, "we should somehow be influenced by our recognition of it," and it should affect how we go about living our lives.[118] Clarke suggests that the propositional core of the religious attitude is that "mentality is eternal."

This religious attitude is consistent with an atheistic panpsychism, as there is no need to posit a Universal Mind, according to Clarke. While

Clarke seeks to demarcate his panpsychist position from that of theism, he is all the more open to recognizing some similarities between his position and Buddhism. According to Buddhism, the Buddha nature is in all things, and this comes close to Clarke's panpsychism.

Clarke further demarcates his position from those panpsychists who suggest that we can properly attribute mentality to the universe as a whole. This is connected to notions such as "World Soul." Can we, in the same way we attribute mentality to singular objects, attribute mentality or a point of view to the universe? The answer to this, according to Clarke, "is obviously no."[119] This is for two reasons.

First, the universe exhibits no unity, not even that of organization or homeostasis. Clarke writes, "Far from exhibiting organization, current astronomy describes it as a chaotic aggregate of galaxies grouped into clusters in which there is continual creation and destruction of both member galaxies and the stars belonging to them."[120]

Second, the universe, in contrast to human consciousness, shows no sign of a single-perspective unity. The universe has no central nervous system, no sense receptors, and "there is no specialization of parts, that could be the basis for attributing some unified perspective" to it.[121]

Clarke is therefore skeptical of attributing a unity to the multitude of minds within the physical universe. This skepticism is most likely a product of his ambition to demarcate atheistic panpsychism from theism. Too much unity of mind could bring his otherwise atheistic metaphysics too close to ontological theism. However, on Clarke's religious view, it is also suggested that an important part of a religious vision, and his own panpsychism, is "an evaluative component" that demands "a certain direction to our lives from us."[122] Clarke further suggests that by "placing priority on the eternal [and thus the mental], we accept the responsibility for maintaining and enhancing natural forms other than those of our own species."[123] Hence, in affirming the eternal nature of the mental, the panpsychist is committed to preserving the diversity of life.

Clarke's panpsychist proposal, however, raises some critical questions. For one thing, given that there is no "unity of mind" in his view, only a multitude of minds with different intentions, interests, and so on, how can one derive a moral system from this ontology? Which mind (or organized unity, as Clarke calls it) is worth preserving? Clarke makes clear that he does not want to ground a moral vision in a universal mind. The problem is that it remains unclear on this position how we should bridge a descriptive statement (that mind is eternal) with a normative statement (that we should preserve the multitudes of mind). Indeed, Clarke seems aware of the difficulty of answering the question "Why be moral?" given this form of panpsychism.[124] Moreover, Clarke seems to suggest that mind has intrinsic worth, given its eternal nature (that mind has always existed in some form). Yet he does not explain how eternity equals intrinsic worth, or how the eternal nature of mind entails that mind has intrinsic value. On the whole, it seems unclear in what sense the affirmation of the eternity of the mental provides any ethical guidelines or justification in the belief that all minds possess intrinsic value. According to Clarke, the moral and practical implications simply follow from panpsychism. Yet he has not sufficiently addressed the issue of how the descriptive statements of panpsychism relate to the normative statements of his ethics.

Contrary to Clarke, Freya Mathews argues that it is appropriate to attribute mentality to the universe as a whole, and that the universe can be considered a Self in its own right. She argues that the universe is a self-realizing system and that "physical reality as a whole can indeed be regarded as an indivisible unity."[125] A self-realizing system, says Mathews, is "defined in system-theoretic terms, as a system with a very special kind of goal, namely its own maintenance and self-perpetuation."[126] Thus, "being a self-realizing system, it possesses reflexivity and to this extent the universe is imbued with a subjectival dimension."[127] Therefore, the universe is not just a unity, it is a subject or a "global self," as Mathews also calls it. Moreover—and Mathews is aware that this is a controversial claim—the Self/Universe might be said to possess awareness. This is not the kind of self-awareness that we observe in higher-level creatures such as animals and humans but a more basic form of awareness. In this way, by ascribing agency to the natural order, Mathews moves her naturalism closer to theism, as we now have teleology and directedness in nature. More precisely, Mathews's view of "the global self" seems to invite a Spinoza-type pantheism.

Is this to push naturalism too far? Is Mathews's approach really naturalistic, or is it merely theism in disguise? It is certainly closer to pantheism, and as such I suggest that it is still compatible with naturalism, if we take a minimal naturalism to be nonsupernaturalistic and committed to the idea of causal closure. That is, Mathews's panpsychist conception of the Divine does not lead to a notion of a being existing outside, or separate from, physical reality. Nor does it lead to any form of extranatural causation given that her conception of the global self is constituted solely by physical elements.

Mathews encounters, nevertheless, another problem. She faces a version of the earlier mentioned combination problem, as she must be able to explain how a multitude of selves can combine in such a way as to constitute a global self, or how many subjects can exist within one great Subject. I would further add to this that if the Many constitutes the unity of the One, then it remains unclear how the agency or will of the One relates to the agency or will of the Many.

Moreover, if Mathews maintains that the One gave rise to the Many, then it immediately raises another question. Namely, why would "a universe whose self-differentiation is integral to its self-realization permit the emergence of finite individuals who possess the potential to disrupt its unfolding?"[128] I highlighted above that the combination problem is particularly pertinent with regard to Mathews's notion of the One and the Many. Here we can see that she also faces the issue of individuation of finite minds within the global mind. These questions remain unsatisfactorily dealt with by Mathews.

Mathews, in emphasizing the self-realizing dimension of physical reality, employs a form of teleology. Not a supernaturalistic form of teleology but perhaps something like Thomas Nagel's immanent teleology, whereby reality is determined not just "by value-free chemistry and physics but also by something else, namely a cosmic predisposition to the formation of life, consciousness, and the value that is inseparable from them."[129] Mathews is clearly not a theist (at least not a classical theist), but her proposal comes very close to a theistic conception of panpsychism.

We saw earlier in this chapter that Stuart Kauffman, in stressing the presence of agency at the quantum level, opens up for the possibility of "cosmic minds." This is a rather exciting shift in Kauffman's thinking, and it creates a space for dialogue concerning the religious potential of a panpsychist ontology. Kauffman's ambition has been to "reinvent the Sacred" and offer a new view of God consistent with science. As was seen above, Kauffman maintains that bursts of consciousness appear due to a measurement event, or an entanglement between particles. Kauffman further suggests that different "sets of entangled variables would be different cosmic minds. And they would be aware—choosing and doing, via the outcomes of measurement—" and yield different actuals.[130] The idea of cosmic minds seems to bring Kauffman's framework closer to theism. It resembles, in many ways, Samuel Alexander's emergent theism, whereby the Deity/God emerges out of the natural order. God is natural and gains a positive ontological status through the configuration between physical constituents and the evolutionary process. This God is real and at the same time a product of natural causation. Interestingly, Kauffman suggests that this panpsychism allows for souls, and even the practice of prayer.[131] This, for Kauffman, is a new spiritual vision and framework for reinventing the sacred, enabled by a panpsychist view of reality.[132]

One must say that Kauffman is more cautious than Mathews when it comes to ascribing agency and unity of consciousness to the totality of the physical. I suggested that Mathews's proposal faces the combination problem of explaining how a multitude of minds (the Many) can combine into a global self (the One), or conversely how the Many can be individuated within the One. Kauffman maintains that his quantum model might offer some relief to the combination problem, as expressed by William James.[133] Kauffman writes, "The wave function for n entangled particles cannot be written as n separate wave functions but only as a single wave function in which we cannot even speak of the n particles as separate."[134] Hence, the particles, with their respective points of view, are no longer separate but are combined into a whole in one wave function. Is this proposal successful? The answer to such question depends on what Kauffman means by "combination." Given that Kauffman talks about a combination of a set of particles, we must expect those particles, with their particular points of view, to survive in the whole. If they do not, we are not talking about combination but about *annihilation* of the parts. That is, when combined they go out of existence and so they bring, ontologically speaking, nothing to the table (and perhaps become epiphenomenal).

Kauffman could respond by saying that despite being grouped into a whole, the particles retain their ontological identity and individuality. Yet if this is his response, then we are back at the original problem of having to explain how a myriad of points of view can coherently combine into a unified point of view, or how a desired "unity of consciousness" is possible. This is not meant to be a defeating critique against Kauffman's panpsychism. It is rather, as I argued with regard to the combination problem, a call for details. Hopefully Kauffman will provide a more thorough treatment of the combination problem in future writings.

Despite some problems and questions in need of further exploration, both Mathews and Kauffman provide panpsychist ontologies that should be of interest for religious naturalists, especially in light of the problems of monistic and pluralistic naturalism. However, we have also seen that Mathews and Kauffman express theistic tendencies. It is now time to explore explicitly theistic understandings of panpsychism.

Theistic Panpsychism: Pierre Teilhard de Chardin and Keith Ward

The Catholic theologian and paleontologist Pierre Teilhard de Chardin proposes a theological vision that opens up a panpsychist understanding of reality. Teilhard's evolutionary theology, which was groundbreaking at the time, seeks to ground God's presence and active involvement firmly within the natural order. The new sciences and our "new awareness of nature's immensities—in the domains of space, time, and organized physical complexity—provides us," according to Teilhard, with a new framework for imagining God's indwelling within the cosmos.[135]

Teilhard critiqued the materialist version of evolution, and suggested instead a Christic interpretation of the direction of evolution and the nature of matter itself. Materialism, argues Teilhard, leads us to think that it is in the lower limits of matter that we can locate the essence, beauty, and riches of the universe. Such materialist philosophy gives us the following principle: "The elements contain in themselves the virtue of the whole: to hold the elements is to possess the whole."[136] Hence, if this principle is true, science would inevitably force us into reductionism. Teilhard wants to replace this materialism with a spiritual interpretation of matter.

For Teilhard, evolution is infused with creative power and teleology. This teleology is expressed in the beginning of one of Teilhard's prayers: "Lord Jesus, you are the centre towards which all things are moving."[137] A key word for Teilhard's evolutionary theology is "convergence," which is the idea that evolution "must culminate ahead in some kind of supreme consciousness."[138] It is by virtue of the "law of complexity-consciousness" that physical constituents have become more complex and given rise to consciousness and self-awareness.[139] This inherent drive of evolution is related to the notion of the Omega Point, the ultimate destiny of the cosmos.[140]

Teilhard's original contribution to Christology is based heavily on the Pauline texts of the Bible, "in which the physical supremacy of Christ over the universe is so magnificently expressed."[141] Christ's incarnation is not merely a past event but relates to the coming into being of the physical universe. Indeed, through the concrete historical incarnation and resurrection of Christ, God's creative power was physically manifested: "So far from the resurrection leaving behind Christ's crucified body, it made that material body the proper and perfect vehicle of Spirit, and mediated God's creative power in bringing the world towards the final unification of matter and spirit."[142]

Christ is both the process within creation as well as its final goal. In this Christocentric view, Christ is seen as identical to Omega. Teilhard suggests that everything "around us is physically 'Christified,' and everything, as we shall see, can become progressively more fully so."[143] For Teilhard it is important to not restrict the salvific work of Christ to the sanctification of human beings.[144] It is with regard to this radical immanence of Christ within the physical that a form of panpsychism begins to emerge, more particularly *pan-Christism*. This is not generic pantheism (God is All), but the affirmation of Christ's radical embodiment in all things (God is All in all).[145]

Christ actively sustains and animates the physical world, and the redeeming work of Christ is possible only to the extent that Christ "could penetrate the stuff of the cosmos" and dissolve "himself in matter."[146] Indeed, for Teilhard, the essential nature of the evolutionary process boils down to "psychical transformation." The (radial) energy that draws the evolutionary process toward increasing complexity is "psychic in nature."[147] Evolution is inherently psychical, and "we are logically forced to assume the existence in rudimentary form . . . of some sort of psyche in every corpuscle, even in those (the mega-molecules and below) whose complexity is of such a low or modest order as to render it (the psyche) imperceptible."[148] Teilhard's Christian theology offers an interesting approach to panpsychism, which seeks to articulate God's active presence through the notion of "pan-Christism."

Another, more recent, theistic understanding of panpsychism comes from Keith Ward. He calls his view "dual-aspect idealism." This is stronger than a general form of panpsychism, which states that all physical stuff contains mental elements or experience. Ward suggests that matter ultimately depends "for its existence on a mind-like reality."[149] Matter, however, is not merely an illusion but has its own form of reality. Matter also allows the "potentialities of the mind to be expressed."[150] Ward further calls this position idealism "because the engine of the process is not the mechanical movements of non-purposive physical entities, but the potentially mind-like reality or realities that are creatively and progressively expressed in the physical cosmos."[151] Ward, similar to Mathews and Teilhard, emphasizes purposive causality and stresses the reality of the cosmic processes that are directed toward certain goals.

It seems that when we look at Mathews, Teilhard, and Ward that panpsychism, construed nontheistically or theistically, invites a sense of teleology. That is, when mind and mindlike properties are taken as basic or fundamental properties, the notion of directedness in nature becomes possible. Ward argues that the strong opposition to teleology in nature, expressed by many scientists and philosophers, is because of the waste and suffering in evolution, and the potential vitalist implications of such teleology. Ward argues convincingly that waste and suffering do not make the teleological structure any less rational or plausible, as there is no reason to suppose that teleological processes are without "blind alleys and eddies." This is no problem as long as the process as a whole leads "to the 'higher' states."¹⁵² Moreover, there is no necessary relationship between this immanent teleology (purposive causality) and vitalism. One can hold to the former without committing to the latter.

Interestingly, and this is not mentioned by Ward, teleology has become slightly more acceptable among naturalists in recent times. We have, for example, the article written by John Hawthorne and Daniel Nolan, "What Would Teleological Causation Be?," which influenced Nagel's view on teleology. In their paper, Hawthorne and Nolan discuss how naturalists usually treat teleological explanations and concepts, which is to reduce them to some underlying reality or treat them metaphorically. Hence, "superficial teleology gives way to an underlying reality that is not fundamentally teleological at all."¹⁵³ In order to explain (away) the appearance of teleology, reductionists employ categories that are not mentalistic at all, such as natural laws. However, building on an Aristotelian framework, Hawthorne and Nolan suggest that teleology might be fundamental, a nonreducible feature, and that it is possible to talk about teleological laws, according to which there is a higher velocity toward certain goals.¹⁵⁴

Returning to Ward, we see that in contrast to Clarke and Mathews, but similar to Teilhard, he links panpsychism/dual-aspect idealism to theism. Given the view of mind as a fundamental feature of reality, and sui generis teleological causation, it is not a great leap to theism. As I have mentioned earlier, there is no necessary relationship between panpsychism and theism, but it can certainly be extended in such direction. Ward maintains more positively that his panpsychist-idealism invites theism. He writes, "Reality itself may be founded not on blind chance or unconscious necessity, but on some form of purposive consciousness."¹⁵⁵ For Ward, theism is not merely an optional add-on; it helps to explanatorily make sense of the mindedness of reality and teleological causation. Given that mind and purpose are among the fundamental components of what is (and given that the universe is directed toward the realization of value and emergence of finite beings), we can conclude that "if there is some ultimate explanation for the existence of the universe, it must include a mind-like element."¹⁵⁶

Theism and Panpsychism

Should, then, a religious naturalist adopt theistic panpsychism as formulated by Ward and Teilhard? I think there are good reasons for a committed panpsychist to consider theism. If a panpsychist ontology turns out to be true, and if there are teleological processes within nature, then we need some explanation for why nature is minded and infused with teleology. The problem with a purely naturalistic interpretation of panpsychism is that the widespreadness of mindlike properties within the natural is left unexplained. Of course, a naturalistic panpsychist might reply, "Mind-qualities are basic and in no need of explanation; bruteness is not a problem." However, it seems as if a panpsychist universe (that is, a universe in which experience is an intrinsic aspect of matter) is in need of some explanation. Unless the naturalistic panpsychist offers good reasons for thinking that experience, or mindlike properties in general, constitutes a *necessary* aspect of matter, that panpsychism would be true in all possible worlds where matter exists, then it seems as if the panpsychist description of the mental-physical relationship is contingent and dependent on the specific configuration of *this* world. Thus the notion of mindlike properties being aspects of the physical should not be accepted as an unexplainable brute fact.

In this way, theism might provide some explanatory relief to the panpsychist depiction of the relationship between the physical and the mental. As Ward suggested, if we seek to explain mind as a fundamental category, then the explanation in question must contain a mindlike element.

A theistic ontology provides a fruitful explanatory framework by which to investigate the phenomenon of fundamental mindlike properties. This point is affirmed by the dualist Charles Taliaferro in his sympathetic treatment of panpsychism. Compared to the materialist or emergent naturalist, the theist can provide a teleological explanation for why the universe exists, why it continues to exist, and why this kind of universe exists. Thus, following Taliaferro, we can conclude that a theistic framework "can provide an account of why there is a panpsychic cosmos and why there should be creatures with minds and different levels of consciousness pervading the cosmos."[157] A minded reality reflects the deeper reality that brought it into existence.

If we take Teilhard's Christocentric view of nature to be true, whereby the physically incarnate Christ directs creation toward greater complexity and unity, then the notion of matter being infused with mind seems quite likely. Indeed, Teilhard's idea regarding the unification of the *many* into *one* becomes very interesting in light of the outlined combination problem. For Teilhard, plurality and unity "is the single problem to which all the physical sciences, all philosophy and all religion fundamentally bring us back."[158] Christ, by virtue of incarnating creation, is the unifying principle, bringing together the One and the Multiple. God is the one "who creates by uniting," as Teilhard puts it.[159] Perhaps the divine influence of Christ, in this Teilhardian view, can be seen as the unifying force, bringing together diverse experiences or qualia into a unified consciousness. This is, of course, a speculative proposal. Nevertheless, it seems as if Teilhard's emphasis on Christ as a unifying force might be of philosophical interest with regard to the combination problem and that of explaining how numerous experiences can combine into full-blown consciousness.

It is worth mentioning that the theisms represented by Ward and Teilhard differ on several points. Religious naturalists do not have to entirely adopt either view, but Ward and Teilhard show us how theism can aid a religiously rich, panpsychist worldview. This is not to force religious naturalists in the direction of ontological theism. Rather, this should be considered an invitation to religious naturalists to consider the ontological resources provided by a theistic framework for explaining a minded reality.

Religious Naturalism and Panpsychism

Let us summarize the implications of the panpsychist framework for religious naturalism. Panpsychism states that experience is widespread in the natural order. Indeed, it can be considered a fundamental feature of both living and nonliving systems. I have also argued that emergence theory, as expressed by religious naturalists, and coupled with semiotics, invites a panpsychist reality.

We have also seen, through Freya Mathews, Keith Ward, and Teilhard de Chardin, the close connection between panpsychism and teleology. Given a "minded nature," it is possible to make sense of the directedness in nature. I have also suggested that panpsychism does not necessitate theism but certainly invites it. Where does this leave religious naturalism? Panpsychism can help religious naturalism to move forward in a number of significant ways:

1. **Metaphysical benefits:** The project of finding a place for higher-level properties in a purely material world has proved exceedingly difficult, and both the reductive and nonreductive form of naturalism encounter severe metaphysical grounding problems. Monistic naturalism ends up reducing, or turning into anomalies, the inevitable presuppositions we make in human practices. Pluralistic naturalism, being more sensitive to such presuppositions, ends up inviting dualism and breaking causal closure. This is an unacceptable implication for naturalism. Panpsychism helps to unify the manifest image with the scientific image by bringing into relationship the mental and physical realm, without denying the fact of

robust consciousness as a later evolutionary development. Hence, mind can be grounded in nature, while taking into consideration the discoveries of science.

2. **Religious benefits:** By virtue of viewing nature as a field of experience, and rejecting the view of nature as dead matter, panpsychism brings with it several important resources for religious and theological consideration. In particular, panpsychism maintains a holistic view of nature that can uphold the religious dimension that several proponents of religious naturalism have associated with emergence theory, the sacredness of nature and its inherent creativity. As panpsychism allows for genuine teleology in nature, it can uphold and indeed make metaphysical sense of the notion of creativity and the ontological claim that there is directedness in nature. Creativity, on panpsychism, can be seen as an immanent teleological process, which calls for a reenchantment of the natural order. I have suggested that such teleology invites theism, but as I see it, there is no necessary relationship between this teleological process and theism.

3. **Ecological benefits:** Panpsychism, by virtue of seeing mind and awareness as pervading the whole of reality, can help us to expand our moral scope, reject moral hierarchy, and include plants, animals, and the ecosystem as a whole within the moral community. We have seen that the philosophical commitment to ecological stability and well-being is crucial for several religious naturalists, and panpsychism is philosophically able to support these important value commitments. This is indeed the ecological promise of panpsychism, which will be further explained in the next chapter.

Conclusions

This chapter has explored the alternative ontology of panpsychism. Panpsychism has for a long time been neglected in philosophical debates but is now gaining renewed attention. It was argued that emergence theory, as formulated by religious naturalists, invites a panpsychist ontology. It was also suggested that the science of biosemiotics gives rise to pansemiosis, which further strengthens the possibility of formulating a panpsychist ontology.

I also responded to some objections to panpsychism, including the infamous combination problem. This combination problem, I argued, was based on a reductionist and antiholistic principle (that the whole is nothing more than the sum of its parts) that we have strong reasons not to accept. I also argued that the objection that panpsychism is not naturalistic (and thus not an option for religious naturalists) is not successful. Moreover, I also responded to Colin McGinn's critique that suggested that weak panpsychism is meaningless.

I outlined several religious approaches to panpsychism, which included nontheistic and theistic understandings. D. S. Clarke, Freya Mathews, and Stuart Kauffman represented nontheistic approaches to panpsychism. It was suggested that Mathews's emphasis on teleology and Kauffman's willingness to entertain the idea of Cosmic Minds bring their respective ontologies significantly closer to theism. I also described Pierre Teilhard de Chardin's and Keith Ward's theistic articulations of panpsychism. It was suggested that while panpsychism does not entail or necessitate theism, the ontology of theism might offer some independent justification for maintaining a panpsychist understanding of the body-mind relationship. In this way, theism helps to avoid leaving the experiential character of the physical unexplained. I also tentatively suggested that theistic panpsychism can offer some explanatory relief to current explorations of the combination problem.

In the following chapter, we will not only review the conclusions of previous chapters but further expand on the promise of panpsychism in the science-religion dialogue with regard to teleology, eco-ethics, and human subjectivity.

9

Concluding Remarks and Looking Ahead

My aim with this book has been to critically evaluate religious naturalism as a position in the dialogue between science and religion. I have outlined and critically evaluated the naturalistic dimension and religious aspect of religious naturalism. Religious naturalists seem to express two different forms of naturalism: monistic naturalism and pluralistic naturalism. The monistic version, a significantly more reductive position, is expressed by both Willem Drees and Charley Hardwick. This monism was critiqued for failing to demarcate itself from sheer reductionism. Although the pluralistic approach fares better at accounting for and providing ontological space for consciousness, it seems to battle another problem, namely dualism. It was argued that emergence theory, and the notion of downward causation in particular, seems to invite dualism, which religious naturalists would deem unacceptable.

It was seen that these naturalists express realistic, antirealistic, and pragmatic understandings of religious discourse, and these three accounts were critically evaluated. I argued for three things. First, although I admire Loyal Rue's defense of religious realism, the evolutionary account of religion seems to undercut his desired realism in the religious sphere. Second, Hardwick's antirealist interpretation of religion is not grounded in his overall ontology, namely physicalism. Therefore, there seems to be a tension between Hardwick's commitment to physicalism and his effort to offer an antirealist take on Christian theology. With regard to Drees's naturalistic religiosity, I argued that his emphasis on limit-questions undermined the plausibility of a naturalistic depiction of the nature and structure of reality. Third, I argued that pragmatic religious realism is compatible with ontological theism and does not necessarily exclude the existence of God or the supernatural. I further argued that this pragmatic strategy seems to epistemically reduce the function of religious language.

In chapters 6, 7, and 8, I discussed alternative ontologies for the project of religious naturalism. Chapter 6 analyzed liberal naturalism, agnostic naturalism, and pragmatic naturalism. Both the liberal and the agnostic versions of naturalism, although admirable attempts to avoid various metaphysical grounding problems, seem to encounter the problem of competing ontologies. Pragmatic naturalism, however, fails to appreciate the seriousness of some philosophical problems, as it reduces such metaphysical debates to mere linguistic conventions. What's more, this form of naturalism undermines the authority of science and thus naturalism itself.

Chapter 7 considered alternative theisms. I began by discussing panentheism, a theistic framework that envisions an intimate connection between God and creation, whereby nature, according to some panentheists, constitutes the body of God. I also suggested that panentheism, being an antisupernaturalistic approach, is closely related to religious naturalism and shares several concerns regarding the effects of traditional religion. However, I suggested that the emergent panentheism proposed by Arthur Peacocke and Philip Clayton, and the process theism advocated by Mark Johnston and David Ray Griffin, ultimately lead to global dualism (pertaining to reality as a whole) as well as local dualism (regarding causation within the natural order). Hence, I concluded that panentheism is not a viable option for a religious naturalist. I evaluated also Fiona Ellis's theistic interpretation of John McDowell's naturalism. I argued that McDowell's expansive naturalism encounters the problem of competing ontologies. Moreover, Ellis's definition of theism was critiqued for its ambiguity and vagueness, which limited the concreteness of her proposal. As such, Ellis's creative proposal was deemed unsuccessful with regard to moving religious naturalism forward.

In chapter 8, I discussed a third alternative ontology, panpsychism. This approach has to a large extent been neglected in theological and philosophical discussions. I considered panpsychism, given its ontological resources, to be the most promising framework for contemporary religious naturalism. In this chapter I will elaborate on some of the benefits of panpsychism for the dialogue between science and religion, especially with regard to teleology, eco-ethics, and the grounding problems regarding subjectivity.

Panpsychism and Teleology

Teleology is one of the great enigmas within philosophy of science and modern biology. Indeed, the concept of "teleology" when associated with "doings" has become an embarrassment for many naturalists. Here I want to reflect on the phenomenon of teleology in nature and its relationship to panpsychism.

Let us first return to Stuart Kauffman, Donald Crosby, and Loyal Rue. These religious naturalists are concerned with teleology in light of the advancements of modern science. As we have seen, they do not want to reject teleology but rather to develop a naturalistically acceptable account of teleology through emergence theory. As was discussed and demonstrated in chapter 3, emergence theory in its current form suffers from several problems. If emergentism is the preferred explanatory strategy for naturalism to accommodate teleology, this project has to be considered rather unsuccessful. Teleology, however, is undeniably real, and any worldview that ends up excluding teleological language must be considered lacking. It is for this reason that Rue, Crosby, and Kauffman come to reject reductive and eliminativist accounts of teleology, and instead propose an emergentist framework for explaining higher-level phenomena.

I have argued for panpsychism as a promising framework for dealing with higher-level phenomena. Indeed, I suggest that what Rue calls a "so that causality" flows naturally from a panpsychist commitment.[1] According to panpsychism, experience is part and parcel of the physical and cannot be ontologically or epistemologically reduced to nonexperiential constituent parts, nor to "efficient causality or to linear causal developments."[2] In this way, panpsychism is able to retain realism regarding teleological notions and directedness in nature. This ontological liberalism allows panpsychism to make sense of "doings" and purpose in the natural realm, as nature is perceived as an experiential domain and not simply a random organization of mindless stuff.

The challenge for the committed naturalist has been to find a home for intrinsic teleology, or purposive causality, in a completely natural world. Given the mechanistic philosophy accompanying materialistic naturalism, teleology and purpose have been ruled out both *within* as well as *for* nature. However, if we reject materialism and mechanistic philosophy, an alternative and richer conception of evolution and natural processes becomes possible. In this view, "modern evolutionary theory, when it is not infected by materialist philosophy, does allow for direction or purpose in the cosmic process."[3] Panpsychism, then, provides for the naturalist a more promising explanatory strategy regarding teleological phenomena. Moreover, the directedness in nature that panpsychism allows for fits well with the "creativity" in nature that several religious naturalists affirm and place at the center of their proposals. In the previous chapter I suggested briefly that panpsychism is consistent with a naturalistic ontology, albeit not a reductive one. This is one significant promise of panpsychism, as it can aid the naturalist in avoiding a potential grounding problem regarding teleology.

In addition, panpsychism, in allowing for teleology within nature, might provide an adequate framework for articulating God's presence within nature.[4] In a theological setting, it seems as if panpsychism, by providing ontological resources for explicating teleology, is relevant for discussions regarding divine action.[5]

I suggested in the previous chapter that there is no necessary relationship between theism and panpsychism. However, panpsychism, in affirming the fundamental nature of the mental and experiential qualities as widespread, seems to align itself with a theistic understanding of reality. We saw how Keith Ward and Pierre Teilhard de Chardin offer frameworks for thinking theistically regarding the fundamental nature of mental qualities. I now want to bring out the teleological aspects of both Ward and Teilhard, and further illustrate how they offer ways for understanding teleology in nature by claiming that the mental is widespread.

Ward suggests, from the perspective of "dual-aspect idealism," that teleological explanations (doings in nature) and the natural realization of the cosmos can be seen as the expression of the initial potency of matter itself.[6] For Ward, the material and mental are "bound together as aspects of a unitary reality."[7] In this teleological view, values are progressively realized in the universe, and the "inner aspect [of matter] is the causal driving force of a cumulative and creative process of increasing organized complexity, generating richer forms of consciousness and purposive causality."[8]

Teilhard traces in a similar way the teleological dimension of the physical to the nature of matter. Matter and spirit, for Teilhard, are not two separate things but two aspects of the same cosmic stuff. The direction of nature is toward ultimate unity, whereby the Omega Point, the universal Christ, is finally realized. Teilhard therefore merges the cosmic, the human, and the Christic through his pan-Christic conception of matter. His theological conception of panpsychism thus offers a unique framework for grounding teleology in nature. Panpsychism offers important ontological resources for grounding teleology, and this should be of interest for naturalists and theists alike.

Panpsychism and Eco-Ethics

Now we shall turn to the area of eco-ethics and consider in what ways panpsychism might benefit current discussions between the discourses of philosophy/theology and ecology. As religious naturalists and other thinkers have pointed out, one of the main reasons for the current ecological situation and the continuous mistreating of nature is faulty metaphysics. That is, metaphysical systems that end up devaluing the rest of nature, or that have introduced rigid dichotomies between humanity and the natural domain, have led and contributed to harmful ways of thinking about and acting toward nature. One important solution to this problem, therefore, is to develop a metaphysical framework that can avoid devaluing nature (making it into an instrument for human purposes), and instead affirm the intrinsic value of the ecosphere.

Panpsychism can avoid some of the significant errors of mechanist philosophies. One philosopher that has considered the promise of panpsychism for the ecological debate is the feminist philosopher Val Plumwood. Similar to my own position, Plumwood argues that mindlike qualities are found throughout nature.[9] There is a strong continuity between mind and nature, and panpsychism rejects the absolute distinction between "the human and the natural spheres."[10] Plumwood rejects in this way the Cartesian notion of the deadness or lifelessness of matter. When the mechanistic and dualistic frameworks are abandoned, another view of nature emerges: "When this framework of discontinuity is discarded, we can see that the major marks of the mental and of what is supposedly distinctive of the human, do not support a picture in which nature is alien but rather one in which nature can be recognised as akin to the human; human difference, like that of other species, appears against an overall background of kinship, forming a web of continuity and difference."[11] The important ethical lesson of panpsychism, if we follow Plumwood, is that it can break down the often-assumed moral hierarchy of the relationship between humanity and the rest of nature, and expand our moral horizon so as to include nonhuman animals and entire ecosystems. Panpsychism, then, offers a conceptual framework that can ground ecological values and avoid ecologically harmful anthropocentric beliefs. Moreover, a panpsychist ontology is significant for the idea of ecologically mindful attitudes (EMA). The panpsychist idea of the experiential, and therefore teleological, nature of physical reality can justify the idea of EMA. That is, when we recognize the inwardness and experiential character of nonhuman organisms, we will be motivated to adopt an EMA. From panpsychism follows a "thou-focused ecological ethics" (as Donald Crosby puts it), and this is highly significant for developing beneficial attitudes toward the natural domain.

This is, of course, a complicated area of discussion, and I am in no way offering panpsychism as a problem-free perspective or a magic quick fix to the current problems pertaining to ecophilosophy. Nevertheless, panpsychism constitutes a promising framework for developing a robust environmental ethics, and my hope is that it will be taken seriously in eco-ethical debates.

Subjectivity and the Promise of Panpsychism

We have seen that both monistic naturalists and pluralistic naturalists struggle to properly ground human subjectivity and agency. It was argued that monistic naturalism, expressed by Charley Hardwick and Willem Drees, is not able to accommodate such beliefs regarding subjectivity. On their monism, our mental life is an anomaly. Consequently, the subjective character of human experience leads to a grounding problem for monistic naturalism. Adopting this form of naturalism comes with a hefty price tag. Pluralistic naturalists, in contrast, seem to invite dualism in their way of grounding mental efficacy, an important part of subjectivity, in naturalism. Panpsychism, by virtue of holding the physical and mental together, can better explain subjectivity, and so avoid such metaphysical grounding problems. Mind is not an anomaly according to panpsychism; mentality is rather ubiquitous and present even in the simplest physical constituents. Philosophers and cognitive scientists recognize the tension between what we often associate with mind (ideas, language, intentionality) and what we associate with our physical body and its biological processes.[12] Panpsychism can escape this Cartesian dichotomy that is the cause of much of the metaphysical anxiety regarding the mind and the self. However, as I have also argued, panpsychism does not entail the absurdity of positing full-blown human consciousness at the fundamental level. Weak panpsychism takes into account the evolved character of the human mind and its organizational complexity. Nevertheless, panpsychism, in viewing the physical as intrinsically experiential, does not end up with the metaphysical puzzlement of having to explain how we get experience from something nonexperiential. Panpsychism may then be able to overcome the metaphysical grounding problem regarding subjectivity that many ontological systems end up with.

In the science-religion dialogue, much effort has been spent on explicating the notion of the "soul" in light of contemporary biology, cognitive science, and neuropsychology.[13] Much of this discussion can be seen as an attempt to overcome reductive physicalism on the one hand, and substance dualism on the other. An emergentist approach is today a very popular, if not the most popular, approach for grounding soul talk within a scientific worldview. Yet emergence theory encounters significant ontological and epistemological problems. I hope to see a shift toward panpsychist thinking within the dialogue between science and religion, as well as attempts to account for the soul and human personhood from the perspective of panpsychism.[14]

Notes

Introduction

1. See Holmes Rolston's (2006: 247–257) treatment of hard and soft naturalism.
2. Crosby 2007: 672.
3. Drees 2010: 110.
4. Stone 1993: 35.
5. Stone 2008: 1.
6. Stone 2008: 1.
7. Cavanaugh 2000: 242; emphasis mine.
8. Drees 2006: 117.
9. Drees 2006: 118.
10. Crosby 2007: 672; emphasis mine.
11. Hogue 2010: 4.
12. Hogue 2010: 5.
13. Hogue 2010: 5.
14. Stone 2008: 13.
15. Hardwick 1996: 22, 75, 260.
16. Hardwick 2003: 114.
17. Goodenough 1998: xvii.
18. Crosby 2002a: 172n. 14.
19. Peters 2012.
20. Peters 2002: 53–58.
21. Kaufman 2004: 42–49.
22. Drees 2006: 121.
23. I explore this issue in M. Leidenhag 2015.
24. Kauffman 2007: 903, 905.
25. Kauffman and Clayton 2006: 515–516.
26. See Rescher 2012; Carroll and Markosian 2010.

Chapter 1

1. Crosby 2007: 672.
2. Stone 2003: 89.
3. Steven Weinberg has famously expressed such a pessimistic view of reality and the possibility of meaning in a natural world. See Weinberg 1993.
4. Stone 1992: 9, 10.
5. Hogue 2010: 5.
6. Stone 1996: 279.
7. Stone 2008.
8. Nadler 2013.
9. Nadler 2013.
10. Nadler 2013.
11. Stone 2008: 21.
12. Arnett 1956: 775.
13. Wieman 2008: 6.
14. Wieman 2008.
15. Drees 2006: 119.
16. Drees 2006.
17. Pihlström 2010: 211.
18. Pihlström 2010: 215.
19. Pihlström 2010: 215.
20. Taylor 2006: 598.
21. Taylor 2006: 599.
22. See, for example, Swinburne 2004.
23. Gregersen 2006: 291.
24. Gregersen 2006: 288.
25. Gregersen 2006: 288–290.
26. Griffin 2004: 83.
27. Griffin 2004: 83.
28. Stone 2008: 9.
29. Jerome Stone writes that religious naturalism involves "the assertion that there seems to be no ontologically distinct and superior realm (such as God, soul or heaven) to ground, explain, or give meaning to this world." Stone 2008: 1.
30. Drees 1997: 531.
31. Goodenough 1998: 28; Kauffman 2008: 203, 229, 231.
32. Murphy 1998: 102.
33. Drees 1996: 183.
34. Drees 1996: 185.
35. Rottschaefer 2001: 467.
36. Stone 2008: 14.
37. Rea 2002: 2, 4.
38. Rea 2002: 5.

39. Hardwick 2003: 112.
40. Hardwick 1996: 51.
41. Hardwick 1996: 34.
42. Kauffman 2008: 7.
43. Kauffman 2008: 8.
44. Cavanaugh 2000: 243.
45. Kaufman 2004: 48.
46. Peters 2002: 38.
47. Peters 2002: 39.
48. Crosby 2002a: xi.
49. Crosby 2002a: 54.
50. Crosby 2002a: 111.
51. Crosby 2002a: 118.
52. Crosby 2002a: 164.
53. Crosby 2008: 43.
54. Crosby 2008: 63.
55. Drees 1996: 267; Drees 1997: 532.
56. Drees 1996: 281.
57. Kauffman 2008: 39.
58. Kauffman 2007: 911.
59. Kauffman 2007: 114.
60. Kauffman 2007: 156, 185.
61. Peters 2008: 84–85.
62. Drees 1996: 92.
63. Drees 1996: 130–131.
64. Drees 1996: 188.
65. Kaufman 2007: 917.
66. Hardwick 1996: 16.
67. Rue 2005: 353.
68. Peters 2002: 102.
69. Peters 2002: 39.
70. Peters 2002: 104.
71. Kaufman 2004: 41.
72. Drees 1998: 617.
73. Kaufman 2004: 53.
74. Peters 2007: 49.
75. Peters 2007: 55.
76. Peters 2007: 60.
77. Kauffman 2008: 284.
78. Kauffman 2008: 276.
79. Drees 1998: 618.
80. Drees 1998: 620.
81. Drees 1998: 621.

82. Rue 2005: 353.
83. Rue 2007: 415.
84. Rue 2005: 366.
85. Crosby 2002a.
86. Goodenough 2000b: 239.
87. Hardwick 1996: 51.
88. Hardwick 2003: 113.
89. Rue 2005: 363.
90. Peters 2002: 74, 91.
91. Goodenough 1998: xvi, 174.
92. Goodenough and Woodruff 2001: 588.
93. Hick 2004a: 348.
94. Peters 2002: 70.
95. Kaufman 2004: 38.
96. Here I draw on Andrew Moore's (2003: 1) characterization of realism.
97. van Fraassen 1980: 7.
98. Drees 1996: 131.
99. Rue 2005: 11–17; M. Leidenhag 2015: 87–99.
100. Hardwick 1996: 33.
101. Insole 2006: 2.
102. Hardwick 2003: 144; Hardwick 1996: 161.
103. Hardwick 1996: 75, 115.
104. Hardwick 1996: 150.
105. Rue 2005: 315.
106. Rue 2005: 316.
107. These are Karl Peters, Stuart Kauffman, Gordon Kaufman, and Ursula Goodenough. Loyal Rue and Donald Crosby represent the realist position, while Willem Drees and Charley Hardwick adopt an antirealist approach.
108. Pihlström 1998: 7, 58.
109. Kauffman 2008: 287–288.

Chapter 2

1. Nagel 1955: 8.
2. Nagel 1955.
3. Nagel 1955: 9.
4. Danto 1967: 448.
5. Kim 2003: 91.
6. Kim 2003: 91–92.
7. See Evan Fales (2007: 118–132) for a discussion regarding the relationship between naturalism and atheism.

8. Quine 1981: 21.
9. Sellars 1922: vii.
10. Raymo 2008: 52.
11. Carl Sagan quoted in Raymo 2008: 55.
12. Rue 2005: 12.
13. Drees 1996: 11.
14. Clayton 2004: 163. I do not consider Clayton to be a religious naturalist. However, his argument resembles that of some religious naturalists. I also employ Clayton in order to outline some of the commitments of pluralistic naturalism, as he is an authoritative figure regarding emergence theory.
15. Drees 2010: 94.
16. Peters 2002: 39.
17. Peters 2002.
18. Drees 1996: 188.
19. Drees 1996: 14.
20. Drees 1996: 14.
21. Drees 1996: 12.
22. Hardwick 1996: 33.
23. Rue 2011: 52; Rue 2005: 14.
24. Peters 2002: 9.
25. Drees 1996: 14.
26. Hardwick 1996: 36.
27. Rue 2011: 102.
28. Churchland and Churchland 1998: 3.
29. Ayer 1936.
30. Kauffman and Clayton 2006: 503.
31. Kauffman and Clayton 2006: 503; Clayton 2004: 5.
32. Clayton 2004: 60.
33. Goodenough and Deacon 2003: 802.
34. Kauffman 2008: 34.
35. Kauffman 2008.
36. Rue 2005: 21, 26, 57.
37. Rue 2000: 100.
38. Rue 2000: 106.
39. Rue 2005: 74.
40. Rue 2011: 85.
41. Rue 2011: 78–79.
42. Kauffman 2008: 39.
43. Kaufman 2004: 92.
44. Kaufman 2004: 93.
45. Kaufman 2007: 924.
46. Peters 2002: 35.

47. Goodenough and Deacon 2003: 813.
48. Crosby 2010: 198.
49. Goodenough and Deacon 2003: 812.
50. Crosby 2002a: 40, 50.
51. Clayton 2006: 4.
52. Niño El-Hani and Emmeche 2000: 242; Clayton 2004: 49.
53. Kauffman and Clayton 2006: 516.
54. Clayton 2004: 62.
55. Kauffman 2008: 7.
56. Crosby 2002a: 47.
57. Crosby 2002a: 47.
58. Crosby 2011: 99.
59. Rue 2005: 316.
60. Rue 2005: 316–317.
61. Kauffman and Clayton 2006: 504.
62. Kauffman 2008: 76.
63. Rue 2011: 58; emphasis in source.
64. Hardwick 1996: 40; emphasis mine.
65. Hardwick 1996: 41.
66. Drees 1996: 189–190.
67. Rue 2005: 12.

Chapter 3

1. Drees 1996: 11.
2. Drees 2002: 18.
3. Drees 2010: 111.
4. Drees 1996: 18.
5. Griffin 2000: 70.
6. Drees 1996: 114.
7. Griffin 2000: 77.
8. Drees 1996: 216.
9. Drees 1996: 176.
10. Drees 1996: 216.
11. Chalmers 1996: 115.
12. Kim 1998: 13.
13. Hardwick 1996: 34.
14. Post 2007: 160.
15. Post 2007: 160.
16. Post 2007: 180–189.
17. Hardwick 1996: 36.

18. Post 2007: 187.
19. Post 2007: 181.
20. Kim 2006: 556.
21. Kim 2006: 556.
22. Post 2007: 181.
23. Drees 1996: 185.
24. Hardwick 1996: 41.
25. *MON1*: The natural world is all there is, and all entities are made up of the same constituents. *MON2*: If X exists, X is either something material or a property of matter. *PSN1*: Teleological language is irreducible and emergent with respect to physics. *PSN2*: Some explanations may require vocabularies that go beyond physics, or the natural sciences in general.
26. Price 2011: 187.
27. Price 2011: 187.
28. Horgan and Timmons 1993: 181.
29. Horgan and Timmons 1993: 182.
30. Horgan and Timmons 1993: 188.
31. Horgan and Timmons 1993: 188.
32. Horgan 1993: 563.
33. This has motivated other naturalists to formulate versions of naturalism more friendly to semantic normativity. See, for example, De Caro and Voltolini 2010.
34. Chalmers 2006: 244.
35. Chalmers 2006: 245.
36. Clayton 2004: 61.
37. Emmeche, Køppe, and Stjernfelt 1997: 100.
38. Kim 1999: 8.
39. Stephan 2006: 488.
40. Bunge 2010: 86.
41. Clayton 2004: 49.
42. Murphy 2006: 228.
43. Kim 2006: 556–557.
44. Gregersen 2006: 280.
45. Monism, here, should not be confused with monistic naturalism.
46. Papineau 2009: 53–65.
47. See Lycan 2013.
48. Kim 2006.
49. Kim 2006: 557–558.
50. Hasker 1999: 60.
51. Hasker 1999: 64.
52. Hasker 1999: 68.
53. Hasker 1999: 190.
54. Sperry 1987: 15, quoted in Hulswit 2006: 269.

55. Hulswit 2006: 266–269.
56. This is what Hulswit (2006: 282–283) argues.
57. Hulswit 2006: 283.
58. Pihlström 2002: 154.
59. Pihlström 2002: 154.
60. Pihlström 2002: 142–148.
61. Pihlström 2002: 158.
62. Pihlström 2002: 148.
63. Pihlström 2002: 153.
64. I elaborate this point further in chapter 6 with regard to Huw Price's pragmatic naturalism.
65. Kauffman 2008: 143.
66. Kaufman 2004: 90.
67. Kaufman 2004: 54.
68. Kaufman 2004: 56.
69. Kaufman 2004: 57–58.
70. Peterson 2006: 693.
71. Rue 2011: 150, 151.
72. Kauffman 2008.
73. Kaufman 2004: 74–75.

Chapter 4

1. Rue 2005: 128.
2. Rue 2005: 129.
3. Rue 2005: 134.
4. Rue 2005: 134.
5. Rue 2005: 136.
6. Rue 2005: 136.
7. Rue 2005: 137.
8. Rue 2005: 140.
9. Rue 2005: 141.
10. Rue 2005: 142.
11. Rue 2005: 142.
12. Rue 2005: 143, 144.
13. Rue 2005: 315.
14. Rue 2005: 317–318.
15. Rue 2005: 324.
16. Rue 2005: 327.
17. Rue 2005: 328.
18. Rue 2005: 131.

19. Rue 2005: 131.
20. Crosby 2002a: 42.
21. Crosby 2002a: 118.
22. Crosby 2002a: 117.
23. Crosby 2008: 27.
24. Crosby 2008: 31.
25. Crosby 2008: 29.
26. Crosby 2008: 146.
27. Crosby 2013: 2.
28. Crosby 2015: 26–27.
29. Crosby 2013: 25.
30. Crosby 2013: 25.
31. Crosby 2013: 45.
32. Hardwick 1987: 26.
33. Hardwick 1996: 4.
34. Hardwick 1996: 4.
35. Hardwick 1996: 82.
36. Hardwick 1987: 25.
37. Hardwick 1987: 18.
38. Hardwick 1987: 18.
39. Hardwick 1996: 255.
40. Hardwick 1996: 257.
41. Hardwick 1996: 257.
42. Hardwick 2001: 196.
43. Hardwick 2001: 65.
44. Hardwick 1996: 109, 119–122.
45. Hardwick 1996: 188.
46. Hardwick 1993: 109.
47. Hardwick 1993: 67.
48. Hardwick 1993: 67.
49. Drees 1996: 250.
50. Drees 1996: 251.
51. Drees 1996: 279.
52. Drees 1996: 276.
53. Drees 1996: 276.
54. Drees 1996: 276.
55. Drees 1996: 278.
56. Drees 2002: 19.
57. Drees 2002: 56.
58. Drees 1996: 114.
59. Drees 1990: 192.
60. Drees 1990: 207.

61. Drees 1990: 197.
62. Drees 1996: 279.
63. Campbell 1995: 14; Diggins 1994: 2.
64. Kolenda 1995: 240.
65. Pihlström 1998: 87.
66. Diggins 1994: 456.
67. Rorty 1997: 6; emphasis mine.
68. Pihlström 1998: 4; emphasis mine.
69. Pihlström 1998: 40.
70. Diggins 1994: 232.
71. Rescher 2012: 22.
72. Zackariasson 2002: 112.
73. Peters 2002: 136.
74. Kaufman 2004: 38.
75. Kaufman 1981: 22.
76. Kaufman 1981: 22.
77. Kaufman 1993: 322.
78. Kaufman 2001: 9.
79. Kaufman 1981: 243.
80. Kaufman 1972: 85.
81. Kaufman 1972: 86.
82. Kaufman 2004: 42.
83. Kaufman 2003: 97.
84. Kaufman 1997: 178.
85. Kaufman 1997: 179.

86. Kaufman writes, "It appears to be qua our development into beings shaped in many respects by *historic-cultural* processes—that is, humanly created, not merely natural biological processes—that we humans have increasingly gained some measure of control over the natural order of which we are part, as well as over the onward movement of history." Kaufman 1997: 178.

87. Kaufman makes the distinction between three different manifestations of the creativity within the natural world: Creativity$_1$ designates the initial coming into being of the physical. Creativity$_2$ refers to later part of universe's history, namely biological evolution. Lastly, Creativity$_3$ refers to "human symbolic activity." Kaufman 2004: 76, 85, 100.

88. Kaufman 2004: 45–46.
89. Peters 2002: vii.
90. Peters 2002: 1.
91. Peters 1998: 314.
92. Peters 1998: 318.
93. Peters 2002.
94. Peters 2002: 35.

95. Peters 2002: 4.
96. Despite Peters's heavy reliance upon Kaufman, there are significant differences between the two. Peters's theology is based on the empiricist tradition in which experience constitutes the philosophical core. Experience is a major guide, indeed the definitive criteria, for judging the validity of religious images. Peters outlines this as one of the major reasons for rejecting a supernaturalistic account of God, as such a God would lie beyond what is accessible to human beings. Kaufman, however, as Peters notes, "seems to shy away from tying theological concepts down to experience." Peters 1993: 200.
97. Peters 1993: 205.
98. Peters 1997: 483.
99. Peters 2002: 83.
100. Kauffman 1995: 16.
101. Kauffman 2008: 11; italics in source.
102. Kauffman 2008: 129.
103. Kauffman 2008: 232.
104. Kauffman 2008: 273.
105. Kauffman 2008: 273.
106. Kauffman 2008: 277.
107. Kauffman 1995: 304.
108. Kauffman 1995: 302.
109. Kauffman 2008: 232.
110. Kauffman 2008: 283.
111. Kauffman 2008: 286.
112. Kauffman 2008: 286.
113. Goodenough 1998: xv.
114. Goodenough 1998: xvii.
115. Goodenough 2000a: 563.
116. Goodenough 1998: 166.
117. Goodenough 2000a: 562.
118. Goodenough 1998: xiv.
119. Goodenough and Deacon 2006: 864.
120. Goodenough and Deacon 2006: 864.
121. Goodenough and Deacon 2006: 867.
122. Goodenough 1998: 72.
123. Goodenough 2001: 392.
124. Goodenough 1998: 11.
125. Goodenough 1998: 167.
126. Goodenough 1998: 102.
127. Goodenough 1998: 86.
128. Goodenough and Woodruff 2001: 585.
129. Goodenough and Woodruff 2001: 593.

130. Goodenough and Woodruff 2001: 588.
131. Goodenough and Woodruff 2001: 591.
132. Goodenough and Deacon 2006: 870.
133. Goodenough 1998: xvi. She writes, "I stand in awe of these religions. I am deeply enmeshed in one of them myself. I have no need to take on the contradictions or immiscibilities between them, any more than I would quarrel with the fact that Scottish bagpipes coexist with Japanese tea ceremonies" (Goodenough 1998: xiv–xv).

Chapter 5

1. Hardwick 1996: 203.
2. Hardwick 1996: 203.
3. Hardwick 1996: 204; emphasis mine.
4. Hardwick 1996: 200.
5. Post 2007: 276.
6. Post 2007: 257.
7. Post 2007: 257.
8. Post 2007: 259.
9. Hardwick 1996: 61.
10. Post 2007: 258.
11. Post 2007: 258.
12. Hardwick 1996: 272.
13. Hardwick 1996: 272.
14. Hardwick 1996: 203.
15. Hardwick 1996: 204.
16. Drees 2002: 14.
17. Drees 1996: 17.
18. Drees 1996: 12.
19. Drees 1996: 188.
20. Hogue 2010: 225.
21. Stenmark 2004: 127.
22. Stenmark 2004: 127.
23. Stenmark 2004: 127.
24. Peters 1993: 200.
25. Peters 1993: 200.
26. Peters 1993: 200.
27. Hick 2004b: 39.
28. Hick 2004a: 140.
29. Hick 2004a: 236.

30. Hick 2004a: 236.
31. Mikael Stenmark (2013: 529–550) has also suggested that religious agnosticism is an option that has been neglected by many religious naturalists.
32. Rescher 2012: 6.
33. See Rescher 2000: 95–105.
34. Rue 2005: 163.
35. Rue 2005: 160.
36. Rue 1994: 84.
37. Rue 1994: 108–124.
38. Rue 1994: 125.
39. Rue 1994: 146.
40. Rue 1994: 150.
41. Rue 1994: 150.
42. Rue 1994: 164.
43. Rue 1994: 203.
44. Rue 1994: 203.
45. Rue 1994: 217–240.
46. Rue 1994: 215.
47. Rue 1994: 194.
48. Rue 1994: 194.
49. Rue 1994: 251.
50. Rue 1994: 274–275.
51. Rue 1994: 279.
52. Rue 1994: 290.
53. Rue 1994: 304.
54. Rue 1994: 306.
55. Hick 2004a: 207.
56. Wallace 1889: 476–477.
57. Hick 2004b: 20.
58. Hick 2004b: 20.
59. Rue 2005: 341–367.
60. Rue 2005: 368.
61. Crosby 2008: ix.
62. Crosby 2008: xi; emphasis in source.
63. Crosby 2008: 25.
64. Crosby 2008: 27.
65. Crosby 2008: 27.
66. Crosby 2008: 70.
67. Crosby 2008: 14.
68. Crosby 2008: 15.
69. See Plantinga 1977.

70. Crosby 2008: 35.
71. Crosby 2013: 24. Crosby recognizes that many organisms in nature do not exhibit sentience, such as "amoebae, bacteria, protists, mites, and other micro-organisms." Crosby 2013: 34.
72. Crosby 2013: 34.
73. Crosby 2013: 53.
74. See Crosby 2002b and Crosby 2010.
75. Crosby 2002a: 84.
76. Crosby 2002a: 88.
77. Crosby 2002a: 89.
78. Crosby 2002a: 89.
79. Crosby 2002a: 89, 87.
80. Crosby 2010: 115.
81. Crosby 2010: 114.
82. Crosby 2008: 58.
83. Crosby 2002a: 100.
84. Wieman 2008.
85. Wieman 2008: 62.
86. Crosby 2008: 63.
87. Crosby 2002a: 163.
88. Crosby 2002a: 161–170.
89. Crosby 2002a: 139; emphasis mine.
90. Crosby 2002a: 166.
91. Crosby 2002a: 166.

Chapter 6

1. De Caro and Voltolini 2010: 75.
2. De Caro and Voltolini 2010: 69–70; emphasis mine.
3. De Caro and Voltolini 2010: 78.
4. Macarthur 2004: 124.
5. De Caro and Voltolini 2010: 79.
6. De Caro and Voltolini 2010: 78.
7. Jackson 2000: 5.
8. Jackson 2000: 5.
9. Jackson 2000: 4.
10. Horgan 1994: 310.
11. See Quine 1963 and Quine 1976. Jaegwon Kim's causal exclusion principle can also be seen as a version of the simplicity argument; see Kim 2005: 17.
12. Jackson 2000: 4.
13. Jaegwon Kim defines supervenience in the following way: "Mental properties strongly supervene on physical/biological properties. That is if any system s

instantiates a mental property M at *t*, there necessarily exists a physical property P such that *s* instantiates P at *t*, and necessarily anything instantiating P at any time instantiates M at that time." Kim 2005: 33.

14. Kim 1999: 21.
15. See, for example, Moreland 2008.
16. Philip Clayton (2004: 108) argues in the following way about the importance of causal efficacy: "One cannot make sense of mental causation except from the standpoint of strong emergence. If the strong emergence interpretation of mental causes is not correct, one should be an epiphenomenalist about mind, that is, one should hold that mind has no effect on the world." In a similar fashion to Clayton, Mark A. Bedau (2008: 175) writes that "emergence is interesting in part because of emergent causal powers. Emergent phenomena without causal powers would be mere epiphenomena."
17. McGinn 2004: 11; italics in source.
18. McGinn 2004: 34.
19. McGinn 2004: 46.
20. McGinn 1989: 350.
21. McGinn 1989: 353; italics in source.
22. McGinn 1993: vii.
23. McGinn 1989: 365.
24. Price 2011: 4.
25. Price 2011: 7.
26. Price 2011: 8.
27. Price 2011: 185.
28. Price 2011: 186.
29. Price 2011: 186.
30. Price 2011: 47.
31. Price 2011: 10.
32. Price 2011: 274–275.
33. Price 2011: 274–275.
34. Price 2011: 136.
35. Price 2011: 138.
36. Williams 2013: 132–133.
37. Tiercelin 2013: 663.
38. Price 2011: 186.
39. Price 2011: 182.

Chapter 7

1. Griffin 2001: 24.
2. Griffin 2000: 11.
3. Griffin 2000: 44.

4. Griffin 2000: 12.
5. Clayton 1997: 169.
6. Peacocke 2007b: 9.
7. Peacocke 2007b: 16.
8. Clayton 1997: 258–259.
9. Johnston 2009: 50–51.
10. Johnston 2009: 52.
11. Johnston 2009: 119–120.
12. For an excellent survey of how to understand the term "causal closure," see Lowe 2000.
13. See Papineau 1993.
14. Kim 1993: 280.
15. Clayton 1997: 181.
16. Clayton 1997: 182.
17. Peacocke 1993: 181; emphasis in source.
18. Clayton 2004: 193.
19. Clayton 2004: 60.
20. Peacocke 2001: 49–50.
21. Peacocke 2007a: 270; emphasis in source.
22. *PON3*: Higher-level Y cannot be reduced to or be replaced by lower-level X. *PON4*: We are epistemologically unable to deduce higher-level entities/properties from low-level physical laws.
23. Griffin 2001: 137.
24. Griffin 2001: 136.
25. Johnston 2009: 119.
26. Griffin 2004: 110. Here we can see that Griffin employs a form of special divine action in his process account of God's incarnation in the person of Jesus.
27. Clayton 1997: 101.
28. Clayton 1997: 183.
29. Ward 2007: 79.
30. Ellis 2014: 4.
31. Ellis 2014: 10.
32. Ellis 2014: 11.
33. Ellis 2014: 17.
34. Ellis 2014: 21.
35. Ellis 2014: 22.
36. Ellis 2014: 34.
37. Ellis 2014: 40.
38. Ellis 2014: 48.
39. Ellis 2014: 51, 52; emphasis in source.
40. Ellis 2014: 66.
41. Ellis 2014: 63.

42. Ellis 2014: 80.
43. Ellis 2014: 86.
44. Ellis 2014: 91.
45. Ellis 2014: 126, 129.
46. Ellis 2014: 162.
47. Ellis 2014: 163.
48. Ellis 2014: 163.
49. Ellis 2014: 115.
50. Ellis 2014: 166.
51. Ellis 2014: 147.
52. Ellis 2014: 178.
53. Ellis 2014: 178.
54. Ellis 2014: 193.
55. Ellis 2014: 151.
56. Ellis 2014: 151.
57. Ellis 2014: 189.
58. Ellis 2014: 189; emphasis in source.
59. McDowell 2004: 92.
60. McDowell 2004: 93.
61. McDowell 2004: 94, 95.
62. McDowell 2004: 97.
63. Ellis 2004: 63.
64. Ellis 2004: 63.
65. Ellis 2004: 71.
66. Ellis 2004: 66.
67. Ellis 2004: 115.
68. Ellis 2004: 95.
69. Ellis 2004: 151.
70. Ellis 2004: 150.
71. Ellis 2004: 151.
72. Ellis 2004: 151.

Chapter 8

1. Searle 1997.
2. McGinn 2006: 93.
3. Clayton 2004: 130.
4. Moreland 2009: 39.
5. Strawson 2006.
6. Chalmers 1996: 128.
7. Rosenberg 2004: 91–113.

8. Nagel 2012: 46.
9. Nagel 2012: 46.
10. Tononi 2004.
11. Tononi and Koch 2015: 11.
12. See Seager (1995: 272–288) for an explanation of the generation problem.
13. Seager 1995: 272.
14. Nagel 1979.
15. Strawson 2006: 7.
16. Strawson 2006: 7.
17. Strawson 2006: 11.
18. Strawson 2006: 12.
19. Strawson 2006: 15.
20. James 1950: 149; emphasis in source.
21. James 1950: 149.
22. Nagel 2012: 56.
23. Nagel 2012: 57.
24. Clarke 2003: 37.
25. Clarke 2004: 12.
26. See Clarke 2003: 30–42.
27. Chalmers 1996: 208.
28. Chalmers 1996: 208.
29. Chalmers 1996: 209.
30. Chalmers 1996: 210.
31. Chalmers 1996: 210.
32. Chalmers 1996: 216.
33. Chalmers 1996: 217.
34. McGinn 1999: 97.
35. Skrbina 2005: 17.
36. Griffin 2004: 78.
37. Rosenberg 2004: 91.
38. Griffin 2001: 120.
39. Griffin 2001: 120.
40. Griffin 2001: 122.
41. Griffin 2001: 122.
42. Kauffman and Clayton 2006: 505.
43. Kauffman and Clayton 2006: 505.
44. Kauffman and Clayton 2006: 517; emphasis mine.
45. Kauffman 2015: 9. Kauffman writes, "I thank Professor Kalevi Kull of the Tartu University Department of Semiotics for convincing me that at just this point, biosemiotics enters."
46. Kauffman 2015: 6.
47. Kauffman 2015: 7.

48. Kauffman 2015: 9.
49. Rue 2011: 57.
50. Rue 2011: 59.
51. Rue 2011: 75.
52. Rue 2011: 81–83.
53. For more on the issue of combinatorial emergence, see M. Leidenhag 2016.
54. Strawson 2006: 12.
55. Griffin 2001: 108.
56. Griffin 2000: 176.
57. Kauffman 2016a: 37.
58. Kauffman 2016a: 37.
59. Kauffman 2016a: 38.
60. Kauffman 2016a: 37.
61. Kauffman 2016a: 39.
62. Kauffman 2016a: 42.
63. Kauffman 2016a: 41.
64. Kauffman 2016a: 42.
65. Kauffman 2016a: 42.
66. Kauffman 2016a: 43.
67. Kauffman 2016a: 45.
68. Goodenough and Deacon 2003: 805.
69. Goodenough 2012: 246.
70. Goodenough 2012: 242.
71. Goodenough 2012: 243.
72. Goodenough 2012: 243.
73. Crosby 2013: 20.
74. Crosby 2013: 21.
75. Crosby 2013: 21.
76. Crosby 2013: 21.
77. Crosby 2013: 19.
78. Crosby 2013: 22.
79. Crosby 2013: 23.
80. Crosby 2013: 24.
81. Crosby 2015: 101.
82. In a recent article, I evaluate Crosby's emergent teleology in relation to Nagel's teleological panpsychism. The main point of this article is to show why Crosby, given the overall problems of an emergentist ontology, should more seriously consider the panpsychist alternative. See M. Leidenhag 2019.
83. Sebeok 2001: 3.
84. Sebeok 2001: 3.
85. Sebeok 2001: 3.
86. Barbieri 2008: 577–596.

87. Emmeche 2004: 321.
88. Emmeche 2004: 319.
89. Emmeche 2004: 319.
90. Emmeche 2004: 325.
91. Emmeche 2004: 326.
92. Emmeche 2004: 327.
93. Chalmers 1996: 293, 297.
94. Seager 1995: 280.
95. Seager 1995: 281.
96. Seager 1995: 281.
97. Skrbina 2005: 263.
98. Goff 2006: 54.
99. Goff 2006: 58.
100. Coleman 2014: 25.
101. Coleman 2014: 26.
102. Coleman 2014: 39–40.
103. Coleman 2014: 39.
104. Rosenberg 2004: 94.
105. Rosenberg 2004: 91.
106. Griffin 2000: 167.
107. Moreland 2008: 133.
108. Clarke 2003: 60.
109. Flanagan 2006: 433.
110. Flanagan 2006: 435.
111. Clarke 2003: 60.
112. Drees 2010: 92.
113. Drees 2010: 92.
114. McGinn 1999: 98.
115. McGinn 1999: 99.
116. Chalmers 1996: 126–127. Chalmers, however, seems to be aware of the problem of specifying the ontology of protophenomenal properties. Because of this problem he thinks that it is more appropriate to consider fundamental properties as phenomenal.
117. Clarke 2003: 145.
118. Clarke 2003: 146.
119. Clarke 2003: 131.
120. Clarke 2003: 131.
121. Clarke 2003: 131.
122. Clarke 2003: 145.
123. Clarke 2003: 169.
124. Clarke 2003: 166.
125. Mathews 2003: 47.

126. Mathews 2003: 47.
127. Mathews 2003: 74.
128. Mathews 2003: 65.
129. Nagel 2012: 123.
130. Kauffman 2016a: 43.
131. Kauffman 2016a: 45.
132. Kauffman 2016a: 46.
133. Kauffman 2016b: 153.
134. Kauffman 2016b: 153.
135. Haught 2002: 540.
136. Teilhard 1965: 27.
137. Teilhard 1961: 76.
138. Teilhard 1999: 183.
139. Haught 2002: 541.
140. Teilhard 1999: 191–194.
141. Teilhard 1965: 54.
142. O'Collins 2011: 521. The idea of final unification is linked to Teilhard's notion of convergence.
143. Teilhard 1965: 59.
144. Teilhard (1965: 58) writes, "In fact, since Christ is omega, he does not restrict his organizing activity simply to one zone of our being—that of sacramental relationships and the 'habitus' of virtues. To enable himself to unite us to him through the highest part of our souls, he has had to undertake the task of making us win through in our entirety, even in our bodies. In consequence, his directing and informing influence runs through the whole range of human works, of material determinisms and cosmic evolutions."
145. Ursula King provides a helpful discussion regarding Teilhard's pantheism and monism. See King 2011: 128.
146. Teilhard 1965: 60.
147. Teilhard 1959: 64.
148. Teilhard 1959: 301–302.
149. Ward 2010: 102.
150. Ward 2010: 92.
151. Ward 2010: 92.
152. Ward 2010: 94.
153. Hawthorne and Nolan 2006: 266.
154. See also Richard Cameron's (2004) article on teleology.
155. Ward 2010: 183.
156. Ward 2010: 184.
157. Taliaferro 2016: 384.
158. Teilhard 1969: 57.
159. Teilhard 1965: 45.

Chapter 9

1. Rue 2011: 42.
2. Crosby 2015: 101.
3. Ward 2010: 98.
4. See Joanna Leidenhag's (2018a) exploration of panpsychism as a resource for divine action.
5. Process theology seeks to explicate models of divine action through panpsychism. However, as seen in chapter 7, process theologians, such as David Ray Griffin, struggle to articulate God's active presence within the natural order due to their overly strong naturalistic commitments (such as the principle of causal closure). Moreover, I suggest that a process theological view of God is deficient in many ways, especially pertaining to the problem of evil. See Pak 2014 and Buckareff 2000.
6. Ward 2010: 100.
7. Ward 2010: 102.
8. Ward 2010: 102.
9. Plumwood 1993: 133.
10. Plumwood 1993: 134.
11. Plumwood 1993: 137.
12. Graves 2008: 24.
13. See Brown, Murphy, and Malony 1998.
14. See Joanna Leidenhag's (2018b) theological articulation of a panpsychist ontology and its implications for Christian understandings of salvation. See also her book *Minding Creation: Theological Panpsychism and the Doctrine of Creation* (2021).

References

Arnett, Willard E. 1956. "Santayana and the Poetic Function of Religion." *The Journal of Philosophy* 22:773–787.
Ayer, Alfred J. 1936. *Language, Truth, and Logic*. London: Victor Gollancz.
Barbieri, Marcello. 2008. "Biosemiotics: A New Understanding of Life." *Naturwisenschaften* 95:577–599.
Bedau, Mark A. 2008. "Downward Causation and Autonomy in Weak Emergence." In *Emergence: Contemporary Readings in Philosophy and Science*, ed. Mark A. Bedau and Paul Humphreys, 155–188. London: MIT Press.
Brown, Warren, Nancey Murphy, and Henry Newton Malony. 1998. *Whatever Happened to the Soul? Scientific and Theological Portraits of Human Nature*. Minneapolis, MN: Fortress Press.
Buckareff, Andrei A. 2000. "Divine Freedom and Creaturely Suffering in Process Theology: A Critical Appraisal." *Sophia* 39:56–69.
Bunge, Mario. 2010. *Matter and Mind: A Philosophical Inquiry*. London. Springer.
Cameron, Richard. 2004. "How to Be a Realist about 'Sui Generis' Teleology Yet Feel at Home in the 21st Century." *The Monist* 87:72–95.
Campbell, James. 1995. *Understanding John Dewey*. Chicago, IL: OpenCourt.
Carroll, John W., and Ned Markosian. 2010. *An Introduction to Metaphysics*. Cambridge: Cambridge University Press.
Cavanaugh, Michael. 2000. "What Is Religious Naturalism? A Preliminary Report of an Ongoing Conversation." *Zygon: Journal of Religion and Science* 35:241–252.
Chalmers, David J. 1996. *The Conscious Mind: In Search of a Fundamental Theory*. New York: Oxford University Press.
Chalmers, David J. 2006. "Strong and Weak Emergence." In *The Re-Emergence of Emergence*, ed. Philip Clayton and Paul Davies, 244–254. New York: Oxford University Press.
Churchland, Paul M., and Patricia S. Churchland. 1998. *On the Contrary: Critical Essays, 1987–1997*. London: MIT Press.
Clarke, D. S. 2003. *Panpsychism and the Religious Attitude*. Albany: State University of New York Press.

Clarke, D. S. 2004. *Panpsychism: Past and Recent Selected Readings*. Albany: State University of New York Press.

Clayton, Philip. 1997. *God and Contemporary Science*. Edinburgh: Edinburgh University Press.

Clayton, Philip. 2000. "Neuroscience, the Person, and God: An Emergentist Account." *Zygon: Journal of Religion and Science* 35:613–652.

Clayton, Philip. 2004. *Mind and Emergence: From Quantum to Consciousness*. New York: Oxford University Press.

Clayton, Philip. 2006. "Conceptual Foundations of Emergence Theory." In *The Re-Emergence of Emergence*, ed. Philip Clayton and Paul Davies, 1–31. New York: Oxford University Press.

Coleman, Sam. 2014. "The Real Combination Problem: Panpsychism, Micro-Subjects, and Emergence." *Erkenntnis* 79:19–44.

Corradini, Antonella. 2010. "The Emergence of Mind: A Dualistic Understanding." In *Causality, Meaningful Complexity and Embodied Cognition*, ed. Arturo Carsetti, 265–273. Dordrecht, Netherlands: Springer Netherlands.

Crane, Tim. 2001. "The Significance of Emergence." In *Physicalism and Its Discontents*, ed. Carl Gillett and Barry Loewer, 207–224. Cambridge: Cambridge University Press.

Crosby, Donald A. 2002a. *A Religion of Nature*. Albany: State University of New York Press.

Crosby, Donald A. 2002b. "Metaphysics and Value." *American Journal of Theology and Philosophy* 23:38–51.

Crosby, Donald A. 2007. "Religious Naturalism." In *The Routledge Companion to the Philosophy of Religion*, ed. Paul Copan and Chad V. Meister, 1145–1162. London: Routledge.

Crosby, Donald A. 2008. *Living with Ambiguity: Religious Naturalism and the Menace of Evil*. Albany: State University of New York Press.

Crosby, Donald A. 2010. "Emergentism, Perspectivism, and Divine Pathos." *American Journal of Theology and Philosophy* 31:196–206.

Crosby, Donald A. 2011. *Faith and Reason: Their Roles in Religious and Secular Life*. Albany: State University of New York Press.

Crosby, Donald A. 2013. *The Thou of Nature: Religious Naturalism and Reverence for Sentient Life*. Albany: State University of New York Press.

Crosby, Donald A. 2015. *Nature as Sacred Ground: A Metaphysics for Religious Naturalism*. Albany: State University of New York Press.

Danto, Arthur C. 1967. "Naturalism." In *The Encyclopedia of Philosophy*, ed. Paul Edwards, 448–450. New York: Macmillan.

De Caro, Mario, and Alberto Voltolini. 2010. "Is Liberal Naturalism Possible?" In *Naturalism and Normativity*, ed. Mario De Caro and David Macarthur, 69–86. New York: Colombia University Press.

Diggins, John. 1994. *The Promise of Pragmatism: Modernism and the Crisis of Knowledge and Authority*. Chicago, IL: University of Chicago Press.

Drees, Willem B. 1990. *Beyond the Big Bang: Quantum Cosmologies and God.* Chicago, IL: Open Court.
Drees, Willem B. 1996. *Religion, Science, and Naturalism.* Cambridge: Cambridge University Press.
Drees, Willem B. 1997. "Naturalism and Religion." *Zygon: Journal of Religion and Science* 32:525–541.
Drees, Willem B. 1998. "Should Religious Naturalists Promote a Naturalistic Religion?" *Zygon: Journal of Science and Religion* 33:617–633.
Drees, Willem B. 2002. *Creation: From Nothing Until Now.* New York: Routledge.
Drees, Willem B. 2006. "Religious Naturalism and Science." In *The Oxford Handbook of Religion and Science*, ed. Philip Clayton and Zachary Simpson, 108–123. New York: Oxford University Press.
Drees, Willem B. 2010. *Religion and Science in Context: A Guide to the Debates.* New York: Routledge.
Ellis, Fiona. 2014. *God, Value, and Nature.* Oxford: Oxford University Press.
Emmeche, Claus. 2004. "Causal Processes, Semiosis, and Consciousness." In *Process Theories: Crossdisciplinary Studies in Dynamic Categories*, ed. Johanna Seibt, 313–336. Dordrecht, Netherlands: Springer Netherlands.
Emmeche, Claus, Simo Køppe, and Frederik Stjernfelt. 1997. "Explaining Emergence: Towards an Ontology of Levels." *Journal for General Philosophy of Science* 28:83–119.
Fales, Evan. 2007. "Naturalism and Physicalism." In *The Cambridge Companion to Atheism*, ed. Michael Martin, 118–134. New York: Cambridge University Press.
Flanagan, Owen. 2006. "Varieties of Naturalism." In *The Oxford Companion of Religion and Science*, ed. Philip Clayton and Zachary Simpson, 430–452. New York: Oxford University Press.
Goff, Philip. 2006. "Experiences Don't Sum." *Journal of Consciousness Studies* 13:53–61.
Goodenough, Ursula. 1998. *The Sacred Depths of Nature.* New York: Oxford University Press.
Goodenough, Ursula. 2000a. "Exploring Resources of Naturalism: Religiopoiesis." *Zygon: Journal of Religion and Science* 35:561–566.
Goodenough, Ursula. 2000b. "Reflections on Scientific and Religious Metaphor." *Zygon: Journal of Religion and Science* 35:233–240.
Goodenough, Ursula. 2001. "Genomes, Gould, and Emergence." *Zygon: Journal of Religion and Science* 36:383–393.
Goodenough, Ursula. 2012. "The Biological Antecedents of Human Suffering." In *The Routledge Companion to Religion and Science*, ed. James W. Haag, Gregory R. Peterson, and Michael L. Spezio, 233–247. New York: Routledge, Taylor and Francis Group.
Goodenough, Ursula, and Paul Woodruff. 2001. "Mindful Virtue, Mindful Reverence." *Zygon: Journal of Religion and Science* 36:585–595.
Goodenough, Ursula, and Terrence W. Deacon. 2003. "From Biology to Consciousness to Morality." *Zygon: Journal of Religion and Science* 38:801–819.

Goodenough, Ursula, and Terrence W. Deacon. 2006. "The Sacred Emergence of Nature." In *The Oxford Handbook of Religion and Science*, ed. Philip Clayton and Zachary Simpson, 853–871. Oxford: Oxford University Press.

Graves, Mark. 2008. *Mind, Brain and the Elusive Soul—Human Systems of Cognitive Science and Religion*. Farnham, UK: Ashgate.

Gregersen, Niels Henrik. 2006. "Emergence: What Is at Stake for Religious Reflection?" In *The Re-Emergence of Emergence*, ed. Philip Clayton and Paul Davies. New York: Oxford University Press.

Griffin, David Ray. 1991. *Evil Revisited: Responses and Reconsideration*. Albany: State University of New York Press.

Griffin, David Ray. 1997. "A Richer or Poorer Naturalism? A Critique of Willem Drees's *Religion Science and Naturalism*." *Zygon: Journal of Religion and Science* 32:593–614.

Griffin, David Ray. 2000. *Religion and Scientific Naturalism: Overcoming the Conflicts*. Albany: State University of New York Press.

Griffin, David Ray. 2001. *Reenchantment without Supernaturalism: A Process Philosophy of Religion*. Ithaca, NY: Cornell University Press.

Griffin, David Ray. 2004. *Two Great Truths: A New Synthesis of Scientific Naturalism and Christian Faith*. Louisville, KY: Westminster John Knox Press.

Hardwick, Charley D. 1987. "Theological Naturalism and the Nature of Religion: On Not Begging the Question." *Zygon: Journal of Religion and Science* 22:21–35.

Hardwick, Charley D. 1993. "The Normative Argument for a Valuational Theism." In *New Essays in Religious Naturalism*, ed. W. Creighton Peden and Larry E. Axel, 99–109. Macon, Georgia: Mercer University Press.

Hardwick, Charley D. 1996. *Events of Grace: Naturalism, Existentialism, and Theology*. Cambridge: Cambridge University Press.

Hardwick, Charley D. 2001. "Foundational Elements in a Naturalistic Ontology." In *Religion in a Pluralistic Age: Proceedings of the Third International Conference on Philosophical Theology*, ed. Donald A. Crosby and Charley D. Hardwick, 189–199. New York: Peter Lang.

Hardwick, Charley D. 2003. "Religious Naturalism Today." *Zygon: Journal of Religion and Science* 39:111–116.

Hasker, William. 1999. *The Emergent Self*. Ithaca, NY: Cornell University Press.

Haught, John F. 2002. "In Search of a God for Evolution: Paul Tillich and Pierre Teilhard de Chardin." *Zygon: Journal of Religion and Science* 37:539–554.

Hawthorne, John, and Daniel Nolan. 2006. "What Would Teleological Causation Be?" In *Metaphysical Essays*, ed. John Hawthorne, 265–283. Oxford: Oxford University Press.

Hick, John. 2004a. *An Interpretation of Religion: Human Responses to the Transcendent*. New Haven, CT: Yale University Press.

Hick, John. 2004b. *The Fifth Dimension: An Exploration of the Spiritual Realm*. Oxford: Oneworld.

Hogue, Michael S. 2010. *The Promise of Religious Naturalism*. Maryland: Rowman & Littlefield.
Horgan, Terence. 1993. "From Supervenience to Superdupervenience: Meeting the Demands of a Material World." *Mind* 102:555–586.
Horgan, Terence. 1994. "Computation and Mental Representation." In *Mental Representation: A Reader*, ed. Stephen P. Stich and Ted A. Warfield, 302–311. Oxford: Basil Blackwell.
Horgan, Terence, and Mark Timmons. 1993. "Metaphysical Naturalism, Semantic Normativity, and Meta-Semantic Irrealism." *Philosophical Issues* 4:180–204.
Hulswit, Menno. 2006. "How Causal Is Downward Causation?" *Journal for General Philosophy of Science* 36:261–287.
Insole, Christopher J. 2006. *The Realist Hope: A Critique of Anti-Realist Approaches in Contemporary Philosophical Theology*. UK: Ashgate.
Jackson, Frank. 2000. *From Metaphysics to Ethics: A Defence of Conceptual Analysis*. New York: Oxford University Press.
James, William. 1950. *The Principles of Psychology*. New York: Dover.
Johnston, Mark. 2009. *Saving God: Religion after Idolatry*. Princeton, NJ: Princeton University Press.
Kauffman, Stuart A. 1995. *At Home in the Universe: The Search for the Laws of Organization and Complexity*. New York: Oxford University Press.
Kauffman, Stuart A. 2007. "Beyond Reductionism: Reinventing the Sacred." *Zygon: Journal of Religion and Science* 42:903–914.
Kauffman, Stuart A. 2008. *Reinventing the Sacred: A New View of Science, Reason, and Religion*. New York: Basic Books.
Kauffman, Stuart A. 2015. "From Physics to Semiotics." In *Issues in Science and Theology: What Is Life?*, ed. Dirk Evers, Michael Fuller, Antje Jackelèn, and Knut-Willy Sæther, 3–19. Cham, Switzerland: Springer.
Kauffman, Stuart A. 2016a. "Cosmic Mind?" *Theology and Science* 14:36–47.
Kauffman, Stuart A. 2016b. *Humanity in a Creative Universe*. New York: Oxford University Press.
Kauffman, Stuart A., and Philip Clayton. 2006. "On Emergence, Agency, and Organization." *Biology and Philosophy* 21:501–521.
Kaufman, Gordon D. 1972. *God the Problem*. Cambridge, MA: Harvard University Press.
Kaufman, Gordon D. 1981. *The Theological Imagination: Constructing the Concept of God*. Philadelphia, PA: Westminster Press.
Kaufman, Gordon D. 1993. *In Face of Mystery: A Constructive Theology*. Cambridge, MA: Harvard University Press.
Kaufman, Gordon D. 1997. "The Epic of Evolution as a Framework for Human Orientation in Life." *Zygon: Journal of Religion and Science* 32:175–188.
Kaufman, Gordon D. 2001. "My Life and My Theological Reflection: Two Central Themes." *American Journal of Theology and Philosophy* 22:3–32.

Kaufman, Gordon D. 2003. "Biohistorical Naturalism and the Symbol of 'God.'" *Zygon: Journal of Religion and Science* 38:95–100.

Kaufman, Gordon D. 2004. *In the Beginning . . . Creativity*. Minneapolis, MN: Fortress Press.

Kaufman, Gordon D. 2007. "A Religious Interpretation of Emergence: Creativity as God." *Zygon: Journal of Religion and Science* 42:915–928.

Kim, Jaegwon. 1993. *Supervenience and Mind: Selected Philosophical Essays*. Cambridge: Cambridge University Press.

Kim, Jaegwon. 1995. "The Myth of Nonreductive Materialism." In *Contemporary Materialism: A Reader*, ed. Paul K. Moser and J. D. Trout, 133–149. London: Routledge.

Kim, Jaegwon. 1998. *Mind in a Physical World: An Essay on the Mind-Body Problem and Mental Causation*. Cambridge, MA: MIT Press.

Kim, Jaegwon. 1999. "Making Sense of Emergence." *Philosophical Studies* 95:3–36.

Kim, Jaegwon. 2003. "The American Origins of Philosophical Naturalism." *Journal of Philosophical Research* 28:83–98.

Kim, Jaegwon. 2005. *Physicalism, or Something Near Enough*. Princeton, NJ: Princeton University Press.

Kim, Jaegwon. 2006. "Emergence: Core Ideas and Issues." *Synthese* 151:547–559.

King, Ursula. 2011. *Teilhard de Chardin and Eastern Religions—Spirituality and Mysticism in an Evolutionary World*. New York: Paulist Press.

Kolenda, Konstantin. 1995. "American Pragmatism and the Humanist Tradition." In *Pragmatism: From Progressivism to Postmodernism*, ed. David Depew and Robert Hollinger, 238–255. London: Praeger.

Leidenhag, Joanna. 2018a. "The Revival of Panpsychism and Its Relevance for the Science-Religion Dialogue." *Theology and Science* 17:90–106.

Leidenhag, Joanna. 2018b. "Saving Panpsychism: A Panpsychist Ontology and Christian Soteriology." In *Being Saved: Explorations in Human Salvation*, ed. Marc Cortez, Joshua R. Farris, and S. Mark Hamilton, 303–325. London: SCM Press.

Leidenhag, Joanna. 2021. *Minding Creation: Theological Panpsychism and the Doctrine of Creation*. London: T&T Clark.

Leidenhag, Mikael. 2014. "Is Panentheism Naturalistic? How Panentheistic Conceptions of Divine Action Imply Dualism." *Forum Philosophicum* 19:209–225.

Leidenhag, Mikael. 2015. "Emergence, Realism, and the Good Life." In *Issues in Science and Theology: What Is Life?*, ed. Dirk Evers, Michael Fuller, Antje Jackelèn, and Knut-Willy Sæther, 87–99. Cham, Switzerland: Springer.

Leidenhag, Mikael. 2016. "From Emergence Theory to Panpsychism: A Critical Evaluation of Nancey Murphy's Non-reductive Physicalism." *Sophia: International Journal of Philosophy and Traditions* 55:381–394.

Leidenhag, Mikael. 2019. "Does Naturalism Make Room for Teleology? The Case of Donald Crosby and Thomas Nagel." *American Journal for Theology and Philosophy* 40:5–19.

Lowe, E. J. 2000. "Causal Closure Principles and Emergentism." *Philosophy* 75:571–585.
Lycan, William G. 2013. "Is Property Dualism Better Off Than Substance Dualism?" *Philosophical Studies* 164:533–542.
Macarthur, David. 2004. "Naturalism and Skepticism." In *Naturalism in Question*, ed. Mario De Caro and David Macarthur, 106–124. London: Harvard University Press.
Mathews, Freya. 2003. *For Love of Matter: A Contemporary Panpsychism*. Albany: State University of New York Press.
McDowell, John. 2004. "Naturalism in the Philosophy of Mind." In *Naturalism in Question*, ed. Mario De Caro and David Macarthur, 91–105. New York: Colombia University Press.
McGinn, Colin. 1988. "Consciousness and Content." *Proceedings of the British Academy* 74:219–239.
McGinn, Colin. 1989. "Can We Solve the Body-Mind Problem?" *Mind* 98:349–366.
McGinn, Colin. 1993. *Problems in Philosophy: The Limits of Inquiry*. Malden, MA: Blackwell.
McGinn, Colin. 1999. *The Mysterious Flame: Conscious Minds in a Material World*. New York: Basic Books.
McGinn, Colin. 2004. *Consciousness and Its Objects*. New York: Oxford University Press.
McGinn, Colin. 2006. "Hard Questions: Comments on Galen Strawson." *Journal of Consciousness Studies* 13:90–99.
Moore, Andrew. 2003. *Realism and Christian Faith: God, Grammar, and Meaning*. Cambridge: Cambridge University Press.
Moreland, J. P. 2008. *Consciousness and the Existence of God: A Theistic Argument*. New York: Routledge.
Moreland, J. P. 2009. *The Recalcitrant Imago Dei: Human Persons and the Failure of Naturalism*. London: SCM Press.
Murphy, Nancey. 1998. "Non-reductive Physicalism: Philosophical Issues." In *Whatever Happened to the Soul? Scientific and Theological Portraits of Human Nature*, ed. Warren Brown, Nancey Murphy, and H. Newton Malony, 127–148. Minneapolis, MN: Fortress Press.
Murphy, Nancey. 2006. "Emergence and Mental Causation." In *The Re-Emergence of Emergence*, ed. Philip Clayton and Paul Davies, 227–243. New York: Oxford University Press.
Nadler, Steven. 2013. "Baruch Spinoza." *The Stanford Encyclopedia of Philosophy*, ed. Edward N. Zalta. Stanford Encyclopedia of Philosophy Archive. Last accessed January 16, 2018. https://plato.stanford.edu/archives/sum2020/entries/spinoza.
Nagel, Ernest. 1955. "Naturalism Reconsidered." *Proceedings and Addresses of the American Philosophical Association* 28:5–17.
Nagel, Thomas. 1979. *Mortal Questions*. Cambridge: Cambridge University Press.
Nagel, Thomas. 2012. *Mind and Cosmos: Why the Materialist Neo-Darwinian Conception of Nature Is Almost Certainly False*. New York: Oxford University Press.

Niño El-Hani, Charbel, and Claus Emmeche. 2000. "On Some Theoretical Grounds for an Organism-Centered Biology: Property Emergence, Supervenience, and Downward Causation." *Theory in Biosciences* 119:234–275.

O'Collins, Gerald, SJ. 2011. "Cosmological Christology: Arthur Peacocke, John Polkinghorne and Pierre Teilhard de Chardin in Dialogue." *New Blackfriars* 93:516–523.

O'Connor, Timothy, and Jonathan Jacobs. 2002. "Emergent Individuals." *The Philosophical Quarterly* 53:540–555.

Pak, Kenneth K. 2014. "Could Process Theodicy Uphold the Generic Idea of God?" *American Journal of Theology and Philosophy* 35:211–228.

Papineau, David. 1993. *Philosophical Naturalism*. Oxford: Blackwell.

Papineau, David. 2009. "The Causal Closure of the Physical and Naturalism." In *The Oxford Handbook of Philosophy of Mind*, ed. Brian P. McLaughlin, Ansgar Beckerman, and Sven Walter, 53–65. Oxford: Clarendon Press.

Peacocke, Arthur. 1993. *Theology for a Scientific Age: Being and Becoming—Natural, Divine, and Human*. Minneapolis, MN: Fortress Press.

Peacocke, Arthur. 2001. *Paths from Science towards God: The End of All Our Exploring*. Oxford: Oneworld.

Peacocke, Arthur. 2006. "Emergence, Mind, and Divine Action: The Hierarchy of the Sciences in Relation to the Human Mind-Brain-Body." In *The Re-Emergence of Emergence*, ed. Philip Clayton and Paul Davies, 257–278. New York: Oxford University Press.

Peacocke, Arthur. 2007a. "Emergent Realities with Causal Efficacy: Some Philosophical and Theological Applications." In *Evolution and Emergence: Systems, Organisms, Persons*, ed. Nancey Murphy and William R. Stoeger, 267–283. New York: Oxford University Press.

Peacocke, Arthur. 2007b. *All That Is: A Naturalistic Faith for the Twenty-First Century*, ed. Philip Clayton. Minneapolis, MN: Fortress Press.

Peters, Karl E. 1993. "Pragmatically Defining the God Concepts of Henry Nelson Wieman and Gordon Kaufman." In *New Essays in Religious Naturalism*, ed. W. Creighton Peden and Larry E. Axel, 199–210. Macon, Georgia: Mercer University Press.

Peters, Karl E. 1997. "Storytellers and Scenario Spinners: Some Reflections on Religion and Science in Light of a Pragmatic, Evolutionary Theory of Knowledge." *Zygon: Journal of Religion and Science* 32:465–489.

Peters, Karl E. 1998. "The Open-Ended Legacy of Ralph Wendell Burhoe." *Zygon: Journal of Religion and Science* 33:313–321.

Peters, Karl E. 2002. *Dancing with the Sacred: Evolution, Ecology, and God*. Harrisburg, PA: Trinity Press International.

Peters, Karl E. 2007. "Toward an Evolutionary Christian Theology." *Zygon: Journal of Religion and Science* 42:49–64.

Peters, Karl E. 2008. *Spiritual Transformations: Science, Religion, and Human Becoming.* Minneapolis, MN: Fortress Press.
Peters, Karl E. 2012. "Human Salvation in an Evolutionary World: An Exploration in Christian Naturalism." *Zygon: Journal of Religion and Science* 47:843–869.
Peterson, Gregory R. 2006. "Species of Emergence." *Zygon: Journal of Religion and Science* 41:689–712.
Pihlström, Sami. 1998. *Pragmatism and Philosophical Anthropology: Understanding Our Human Life in a Human World.* New York: Peter Lang.
Pihlström, Sami. 2002. "The Re-Emergence of the Emergence Debate." *Principia* 6:133–181.
Pihlström, Sami. 2010. "Dewey and Pragmatic Religious Naturalism." In *The Cambridge Companion to Dewey*, ed. Molly Cochran, 211–241. Cambridge: Cambridge University Press.
Plantinga, Alvin. 1977. *God, Freedom, and Evil.* Grand Rapids, MI: Wm. B. Eerdmans.
Plantinga, Alvin. 2006. "Against Materialism." *Faith and Philosophy* 23:3–32.
Plumwood, Val. 1993. *Feminism and the Mastery of Nature.* London: Routledge.
Post, John F. 2007. *The Faces of Existence: An Essay in Non-reductive Metaphysics.* London: Cornell University Press.
Price, Huw. 2011. *Naturalism without Mirrors.* New York. Oxford University Press.
Quine, Willard van Orman. 1963. "On Simple Theories of a Complex World." *Synthese* 15:103–106.
Quine, Willard van Orman. 1976. "Posits and Reality." In *The Ways of Paradox and Other Essays*, ed. Willard Van Orman Quine, 246–254. Cambridge, MA: Harvard University Press.
Quine, Willard van Orman. 1981. *Theories and Things.* Cambridge, MA: Harvard University Press.
Raymo, Chet. 2008. *When God Is Gone Everything Is Holy: The Making of a Religious Naturalist.* Notre Dame, IN: Sorin Books.
Rea, Michael C. 2002. *World without Design: The Ontological Consequences of Naturalism.* Oxford: Clarendon Press.
Rescher, Nicholas. 2000. "God's Place in Philosophy (Non in Philosophia Recurrere Est ad Deum)." *Philosophy and Theology* 12:95–105.
Rescher, Nicholas. 2012. *Pragmatism: The Restoration of Its Scientific Roots.* New Brunswick, NJ: Transaction.
Rolston, Holmes. 2006. *Science and Religion: A Critical Survey.* Philadelphia: Templeton Foundation Press.
Rorty, Richard. 1997. "Religious Faith, Intellectual Responsibility, and Romance." In *Pragmatism, Neo-Pragmatism, and Religion: Conversations with Richard Rorty*, ed. Donald A. Crosby and Charley D. Hardwick, 3–21. New York: Peter Lang.
Rosenberg, Gregg. 2004. *A Place for Consciousness: Probing the Deep Structure of the Natural World.* New York: Oxford University Press.

Rottschaefer, William A. 2001. "Discerning the Limits of Religious Naturalism." *Zygon: Journal of Religion and Science* 36:467–475.
Rue, Loyal. 1994. *By The Grace of Guile: The Role of Deception in Natural History and Human Affairs*. New York: Oxford University Press.
Rue, Loyal. 2000. *Everybody's Story: Wising Up to the Epic of Evolution*. Albany: State University of New York Press.
Rue, Loyal. 2005. *Religion Is Not about God: How Spiritual Traditions Nurture Our Biological Nature and What to Expect When They Fail*. New Brunswick, NJ: Rutgers University Press.
Rue, Loyal. 2007. "Religious Naturalism—Where Does It Lead?" *Zygon: Journal of Religion and Science* 42:409–422.
Rue, Loyal. 2011. *Nature Is Enough: Religious Naturalism and the Meaning of Life*. Albany: State University of New York Press.
Seager, William. 1995. "Consciousness, Information, and Panpsychism." *Journal of Consciousness Studies* 2:272–288.
Searle, John R. 1997. "Consciousness and the Philosophers." *The New York Review of Books,* March 6. Accessed November 19, 2018. https://www.nybooks.com/articles/1997/03/06/consciousness-the-philosophers.
Searle, John R. 2001. *Rationality in Action*. London: MIT Press.
Searle, John R. 2008. "Reductionism and the Irreducibility of Consciousness." In *Emergence: Contemporary Readings in Philosophy and Science*, ed. Mark A. Bedau and Paul Humphreys, 69–80. London: MIT Press.
Sebeok, Thomas A. 2001. *Signs: An Introduction to Semiotics*. Canada: University of Toronto Press.
Sellars, Roy Wood. 1922. *Evolutionary Naturalism*. Chicago, IL: Open Court.
Skrbina, David. 2005. *Panpsychism in the West*. London: MIT Press.
Sperry, Roger W. 1987. "Consciousness and Causality." In *The Oxford Companion to the Mind*, ed. Richard L. Gregory, 164–166. Oxford: Oxford University Press.
Stenmark, Mikael. 2004. *How to Relate Science and Religion: A Multidimensional Model*. Grand Rapids, MI: Wm. B. Eerdmans.
Stenmark, Mikael. 2013. "Religious Naturalism and Its Rivals." *Religious Studies* 49:529–550.
Stephan, Achim. 2006. "The Dual Role of 'Emergence' in the Philosophy of Mind and in Cognitive Science." *Synthese* 151:485–498.
Stone, Jerome A. 1992. *The Minimalist Vision of Transcendence: A Naturalist Philosophy of Religion*. Albany: State University of New York Press.
Stone, Jerome A. 1993. "The Viability of Religious Naturalism." *American Journal of Theology and Philosophy* 14:35–42.
Stone, Jerome A. 1996. "Caring for the Web of Life: Towards a Public Ecotheology." In *Religious Experience and Ecological Responsibility*, ed. Donald A. Crosby and Charley D. Hardwick, 277–285. New York: Peter Lang.
Stone, Jerome A. 2003. "Varieties of Religious Naturalism." *Zygon: Journal of Religion and Science* 38:89–93.

Stone, Jerome A. 2008. *Religious Naturalism Today: The Rebirth of a Forgotten Alternative*. Albany: State University of New York Press.
Stone, Jerome A. 2017. *Sacred Nature: The Environmental Potential of Religious Naturalism*. Routledge: London.
Strawson, Galen. 2006. "Realistic Monism: Why Physicalism entails Panpsychism." *Journal of Consciousness Studies* 13:3–31.
Swinburne, Richard. 2004. *The Existence of God*. 2nd ed. New York: Oxford University Press.
Swinburne, Richard. 2007. "From Mental/Physical Identity to Substance Dualism." In *Persons: Human and Divine*, ed. Peter van Inwagen and Dean Zimmerman, 142–165. Oxford: Oxford University Press.
Taliaferro, Charles. 2016. "Dualism and Panpsychism." In *Panpsychism: Contemporary Perspectives*, ed. Godehard Brüntrup and Ludwig Jaskolla, 369–386. Oxford: Oxford University Press.
Taylor, Bron. 2006. "Religion and Environmentalism in American and Beyond." In *The Oxford Handbook of Religion and Ecology*, ed. Robert S. Gottlieb, 588–612. Oxford: Oxford University Press.
Teilhard de Chardin, Pierre. 1959. *The Phenomenon of Man*. New York: Harper and Row.
Teilhard de Chardin, Pierre. 1961. *Hymn of the Universe*. London: William Collins Sons.
Teilhard de Chardin, Pierre. 1965. *Science and Christ*. London: William Collins Sons.
Teilhard de Chardin, Pierre. 1999. *The Human Phenomenon*. Brighton, UK: Sussex Academic Press.
Tiercelin, Claudine. 2013. "No Pragmatism without Realism." *Metascience* 22:659–655.
Tononi, Giulio. 2004. "An Information Integration Theory of Consciousness." *BMC Neuroscience* 5:42.
Tononi, Giulio, and Christof Koch. 2015. "Consciousness: Here, There, and Everywhere?" *Philosophical Transactions B* 370:1–18.
van Fraassen, Bas C. 1980. *The Scientific Image*. Oxford: Clarendon Press.
Wallace, Alfred. 1889. *Darwinism: An Exposition of the Theory of Natural Selection with Some of Its Applications*. New York: Macmillan.
Ward, Keith. 2007. *Divine Action: Examining God's Role in an Open and Emergent Universe*. West Conshohocken, PA: Templeton Foundation Press.
Ward, Keith. 2010. *More Than Matter: Is There More to Life than Molecules* Grand Rapids, MI: William B. Eerdmans.
Weinberg, Steven. 1993. *The First Three Minutes: A Modern View of the Origin of the Universe*. New York: Basic Books.
Wieman, Henry Nelson. 2008. *The Source of Human Good*. Eugene, OR: Wipf and Stock.
Williams, Michael. 2013. "How Pragmatists Can Be Local Expressivists." In *Expressivism, Pragmatism and Representationalism*, ed. Huw Price, 128–144. Cambridge: Cambridge University Press.

Zackariasson, Ulf. 2002. *Forces by Which We Live: Religion and Religious Experience from the Perspective of a Pragmatic Philosophical Anthropology.* Stockholm, Sweden: Almquist and Wiksell.

Name Index

Alexander, Samuel, 1, 17, 71, 209
Ayer, A. J., 49

Bedau, Mark A., 241n16
Bohm, David, 192
Bultmann, Rudolf, 89–90
Burhoe, Ralph Wendell, 100–101

Cavanaugh, Michael, 4, 5, 21
Chalmers, David, 62, 67, 183, 186–187, 199, 206, 246n116
Churchland, Patricia, 49
Churchland, Paul, 49, 64
Clarke, D. S., 203–204, 206–208, 213, 217
Clayton, Philip, 17, 45, 50, 53, 57, 68, 161–162, 163–166, 167, 169, 170, 179, 182, 220, 231n14, 241n16
Coleman, Sam, 201–202, 203
Crosby, Donald A., 2–3, 5, 6, 14, 22–23, 30, 31, 52–53, 54, 86–89, 106, 109, 125–134, 135, 190, 195–196, 221, 223, 230n107, 240n71, 245n82

Danto, Arthur C., 42
Darwin, Charles, 29, 100, 195
Dawkins, Richard, 11
Deacon, Terrence, 191

De Caro, Mario, 137–139, 140, 141, 142, 143, 146, 172, 181
Democritus, 41
Dennett, Daniel, 12
Dewey, John, 13, 14, 94, 167
Dombrowski, Donald, 130
Drees, Willem B., 2, 3, 4–5, 7, 14, 15, 16–17, 19, 20, 21, 23–24, 25–26, 29, 30, 36, 44–45, 47, 48, 56, 57, 59, 60–62, 63, 64, 66, 80, 87, 89, 92–94, 107, 109, 114–116, 120, 134, 143, 184, 204, 219, 224, 230n107

Ellis, Fiona, 9, 159, 160, 171–179, 180, 182, 220
Emmeche, Claus, 68, 73, 197–199

Fales, Evan, 230n7
Ferré, Frederick, 130
Flanagan, Owen, 204

Galilei, Galileo, 24
Geertz, Clifford, 99
Goff, Philip, 201
Goodenough, Ursula, 2, 5, 6, 30, 31, 34, 50, 52, 94, 97, 103–106, 107, 109, 116, 135, 190, 194–195, 196, 203, 230n107, 238n133
Gregersen, Niels Henrik, 16, 17, 71

Index

Griffin, David Ray, 18, 61, 62, 161–162, 167–168, 170, 179, 182, 188–189, 192, 203, 220, 242n26, 248n5
Griffin, Donald, 195

Hardwick, Charley D., 2, 5–6, 19, 20–21, 24, 26, 29, 32–33, 36–37, 47, 48, 55–56, 57, 59, 60, 63–64, 66, 80, 87, 89–92, 93, 94, 107, 109–113, 114, 120, 134, 184, 219, 224, 230n107
Harris, Sam, 11
Hartshorne, Charles, 186
Hasker, William, 72–73
Hawthorne, John, 213
Heidegger, Martin, 111
Hick, John, 119, 123, 124, 135
Hiley, Basil, 192
Hogue, Michael S., 5, 116
Horgan, Terry, 65, 142–143
Hulswit, Menno, 73–74

Jackson, Frank, 142, 143
James, William, 185, 200, 210
Jesus, 168, 175, 211–212, 214–215, 242n26, 247n144
Johnston, Mark, 162, 167, 168, 170, 179, 220

Kant, Immanuel, 75, 98
Kaufman, Gordon D., 2, 3, 5, 6, 14, 22, 26, 27–28, 29, 30, 33–34, 35, 36, 37–38, 52, 76, 78, 79, 94, 97–100, 101, 102, 103, 106, 107, 109, 116, 119, 134, 167, 230n107, 237n96
Kauffman, Stuart, 2, 3, 6, 7, 14, 21, 24, 29–30, 33, 38, 50, 51–52, 53–54, 55, 56, 68, 76, 78, 79, 94, 97, 102–103, 106, 107, 109, 116, 134, 148, 149, 189–194, 195, 196, 206, 209–210, 217, 221, 230n107, 244n45
Kim, Jaegwon, 63–64, 70, 72, 74, 145, 163, 240n11, 240–41n13
King, Ursula, 247n145
Kiwi the cat, 128–129
Koch, Christof, 183
Køppe, Simo, 68, 73
Kull, Kalevi, 190, 191, 244n45

Laplace, Pierre-Simon, 24
Leidenhag, Joanna, 248n4, 248n14
Leidenhag, Mikael, 245n82
Levinas, Emmanuel, 174–175, 178
Loomer, Bernard, 133
Lucretius, 41

Macarthur, David, 138–139, 140, 141, 142, 143, 146, 172, 181
Mathews, Freya, 206, 208–209, 210, 212, 213, 215, 217
McDowell, John, 75, 160, 171, 172, 173–174, 176–177, 179, 180, 182, 220
McGinn, Colin, 146–151, 156, 181, 182, 205–206, 217
Monod, Jacques, 191
Moreland, J. P., 182, 203
Morgan, Lloyd, 71
Murphy, Nancey, 69

Nagel, Ernest, 42
Nagel, Thomas, 183, 184, 185–186, 196, 209, 213, 245n82
Newton, Isaac, 24
Nolan, Daniel, 213

O'Connor, Timothy, 145

Papineau, David, 71, 163
Peacocke, Arthur, 17, 162, 163–164, 165–166, 167, 170, 179, 182, 220

Peirce, Charles Sanders, 101, 118, 119
Peters, Karl E., 2, 3, 6, 14, 22, 24–25, 27, 28–29, 30, 33, 34, 35, 46, 47, 52, 78, 94, 97, 100–102, 103, 106, 107, 109, 116, 118–119, 134, 230n107, 237n96
Peterson, Gregory, 78
Pihlström, Sami, 14, 74–76
Plantinga, Alvin, 128
Plumwood, Val, 223
Polkinghorne, John, 25
Post, John, 21, 63–64, 90, 111–113, 114
Price, Huw, 65, 151–156, 157, 181, 234n64
Putnam, Hilary, 75

Quine, Willard Van Orman, 42–43, 143

Railton, Peter, 173
Raymo, Chet, 43–44
Rea, Michael C., 20
Rescher, Nicholas, 96, 119–120
Rolston, Holmes, 227n1
Rorty, Richard, 95
Rosenberg, Gregg, 183, 202–203, 205
Rottschaefer, William A., 20
Rue, Loyal, 2, 3, 5, 6, 27, 30–31, 34, 36, 37, 44, 47, 48, 51, 54, 55, 56, 57, 78, 83–86, 87, 89, 92, 106, 107, 109, 120–124, 132, 135, 143, 189, 191, 192, 195, 196, 219, 221, 230n107

Sagan, Carl, 44
Santayana, George, 1, 13, 14
Seager, William, 192, 200

Searle, John, 182
Sellars, Roy Wood, 43, 47
Skrbina, David, 188, 200
Sperry, Roger W., 74
Spinoza, Baruch, 12–13
Stenmark, Mikael, 117, 239n31
Stjernfelt, Frederik, 68, 73
Stone, Jerome A., 1, 4, 5, 12, 13, 18, 20, 228n29
Strawson, Galen, 183, 184–185, 192

Taliaferro, Charles, 214
Teilhard de Chardin, Pierre, 206, 210–212, 213, 214–215, 217, 222, 247n142, 247n144, 247n145
Thompson, Evan, 88, 129, 195–196
Tillich, Paul, 167
Timmons, Mark, 65
Tononi, Giulio, 183

Uexküll, Jakob von, 198

Van Fraassen, Bas, 35
Voltolini, Alberto, 137–139, 140, 141, 142, 143, 146, 181
Von Wright, Georg Henrik, 75

Wallace, Alfred Russell, 124
Ward, Keith, 170–171, 206, 212–213, 214, 215, 217, 222
Weinberg, Steven, 228n3
White, Lynn, Jr., 27
Whitehead, Alfred North, 87, 186
Wieman, Henry Nelson, 13–14, 133, 134
Williams, Michael, 155
Wittgenstein, Ludwig, 75, 119

Subject Index

abstract objects, 138, 140
agency, 7, 24, 61–62, 79, 102, 195–197, 204, 208–210, 224
  molecular, 190–194
aggregational societies, 188–189. *See also* compound individuals
agnosticism. *See* naturalism, agnostic; ontology, agnosticism
animism, 15
anthropocentricity, 27, 28, 30, 99, 223
antirealism, 11, 35, 39, 96, 120, 128, 156
  religious, 8–9, 36–37, 38, 39, 89–94, 106, 107, 109, 159, 219, 230n107
antireductionism, 59, 60–66, 154. *See also* nonreductionism; reductionism
antisupernaturalism, 204, 220. *See also* supernaturalism
atheism, 11–12, 36, 114, 161, 174, 175, 230n7
autopoiesis, 195

biohistory, 33, 98–100. *See also* naturalism, biohistorical
biological diversity, 88, 123, 126, 129, 207
biology, 43, 50–51, 53, 79, 111, 191, 197, 199, 220, 224. *See also* evolution

biosemiotics, 190, 196–199, 204, 216, 244n45. *See also* pansemiosis; semiotics
Buddhism, 207

causation, 47–48, 60, 62, 67–68, 138–139, 140, 145, 154, 161–171, 178–179, 189, 212–213, 221–222, 241n16
  causal closure, 16, 42, 71–73, 75, 80, 154, 160, 162–163, 165–166, 168, 170–171, 182, 208, 215, 248n5
  causal exclusion principle, 240n11
  downward, 53, 57, 59, 67, 69–70, 71–76, 77–78, 80, 145, 166, 179, 219
  emergent, 74
  mental, 241n16
  "space of causes," 176–177
Christianity, 5, 6, 9, 14, 26, 27, 28–29, 31, 37, 86, 89–91, 98, 100, 109–113, 134, 248n14. *See also* theism, Christian
Christology, 98, 211–212, 214–215, 222, 247n144
  incarnation, 168, 211, 242n26
  pan-Christism, 212, 222
  resurrection, 211
cognition, 188, 196, 202–203
compound individuals, 188–189, 203. *See also* aggregational societies

265

consciousness, 2, 19, 49, 61–63, 115, 147, 149–150, 162, 181–187, 188, 189–199, 202, 205, 206, 209–210, 211, 214–215, 216, 224. *See also* mentality
   integrated information theory of, 183
   macroconsciousness, 200, 201
   microconsciousness, 200, 201
constructivism, 38, 86
cosmology, 28, 33, 83, 104, 114
creatio ex nihilo, 18, 167
creativity, 3, 13–14, 22, 28, 29, 34, 38, 52, 76, 78, 99–100, 101, 216, 221, 236n87

deism, 46, 163, 164–165, 169, 178
deception, 121–123
demythologization, 89–90
determination, 63, 111
determinism, 19, 47, 48, 61–62, 75, 78, 112, 161, 209
divine action, 9, 25, 26, 60, 92–93, 160, 163–165, 166, 168–170, 171, 178–179, 222, 248n4, 248n5. *See also* providence
   influence, 161, 164, 168–171 (*see also* persuasion)
   interventionism, 18, 42, 115, 159, 160, 161, 163, 167, 168, 178, 181–182
   noninterventionism, 60, 114
   special divine action, 164, 165, 242n26
dualism, 18, 26, 42, 62, 115, 141, 145, 146, 150, 160–161, 162, 187, 219
   emergent, 72–73, 142
   global, 167, 182, 220
   local, 167, 182, 220
   mind-body, 47, 62, 146–148
   naturalistic, 187
   and panentheism, 163–170

ecology, 5–6, 12, 27–28, 30, 33–35, 38, 97, 98–99, 100, 101, 103–104, 107, 116–118, 120, 123
eco-ethics, 12, 15, 105–106, 129, 220. *See also* morality
   and panpsychism, 216, 217, 222–224
ecologically mindful attitudes, 32, 34–35, 39, 97, 103, 107, 116, 120, 223
ecotheology, 3, 12, 118–119
eliminativism, 19, 55, 60, 62, 63, 64, 80, 151, 184, 201
emergence theory, 17, 49–53, 67–70, 71, 76–78, 79–81, 144–145, 162, 163–167, 179, 185, 186, 196, 199, 205, 216, 219, 221, 224–225, 231n14, 241n16, 245n82. *See also* causation, emergent; epistemology, emergent; naturalism, emergent; ontology, novelty; pluralism, emergent
   combinatorial, 191–192, 245n53
   and panentheism, 9, 17, 18, 220
   and panpsychism, 9, 184–185, 189–199
   strong, 7, 51, 52–53, 57, 57–58, 67, 74, 77, 78, 145–146, 166, 170, 186, 189, 241n16
   and teleology, 51, 245n82
   weak, 51–52, 67, 75, 189, 197, 199
emergent properties, 19, 67–70, 72–73, 79–81, 148–149
epiphenomenalism, 65, 72, 77, 78, 145, 149, 167, 210, 241n16
epistemology, 20–21, 112, 115–116, 169–170. *See also* reductionism, epistemic
   epistemic simplicity, 141, 143, 146
   evidentialism, 48
   monistic, 21, 38, 48, 56, 57, 109–110
   pluralistic, 21, 38, 54
ethics. *See* eco-ethics; morality

evil, 23, 87–88, 125–134, 248n5. *See also* pain; suffering
evolution, 22, 26, 52, 55, 92, 99, 124, 195–196, 221. *See also* biology
epic of evolution, 3, 4, 51, 100
existentialism, 87, 89–90, 107, 110, 111, 113, 124
expressivism
  global, 152, 154, 155
  local, 155

facts and values, 111–113
fictionalism, 123–124, 151, 153, 156
foundationalism, 94–95
free will, 7, 49, 127–128, 193
functionalism, 13, 14, 16, 87, 92, 96, 120–124, 156

God, 16–18, 237n96
  as cause among causes, 178–179
  as creativity, 3, 7, 13–14, 22, 28, 29, 34, 38, 79, 99–100, 101, 107
  as Creator, 7, 17, 22, 26, 178
  and God-talk, 3, 112, 153, 170, 178–179
  moral goodness of, 127–128
  as mystery, 76
  as natural processes, 3, 22, 33, 34, 79, 100, 101
  as Nature, 12–13, 29
  as totality of world, 133
  and values, 171–179
grace, 11
  events of, 2, 24–25, 91, 130

humanities, the, 21, 54, 59, 173

idealism, 37
  dual-aspect idealism 212, 213, 222
instrumentalism, 87, 91, 93, 129–130
intentionality, 88, 129, 148, 150, 165, 168, 196, 224

Islam, 86

Judaism, 14, 37, 86

laws of nature. *See* natural laws
limit-questions, 2, 16, 21, 23–24, 60, 93, 94, 114–116, 134, 219

materialism, 19, 36, 41, 47, 57, 60, 61, 62, 64, 73, 124, 152, 153, 161, 185–186, 211, 214, 221. *See also* physicalism
matter. *See* dualism, mind–body; materialism; physicalism
mechanistic philosophy, 203, 221, 223
mentality. *See also* consciousness; dualism, mind-body
  as eternal, 206–208
  protomentality, 186, 205
mental revelation, 167–168, 180
metaphysical grounding problems, 7–8, 43, 58, 59–81, 113, 137, 142, 154, 156, 177, 215, 220, 224
microorganisms, 192, 194–195, 203, 240n71
mind. *See also* mentality
  cosmic minds, 193–194, 209, 217
  philosophy of, 62, 147
miracles. *See* psychological miracles
monism. *See* ontology, monism
morality. *See* eco-ethics
mystery, 3, 66, 68, 71, 76–78, 79, 93, 99, 102–103, 105, 114, 116, 140, 142, 167, 179, 184
myth, 30, 31, 37, 39, 83–86, 89, 92, 121, 122, 123, 124

naturalism, 18–21, 208, 220, 221, 224, 230n7. *See also* antinaturalism; religious naturalism
  agnostic, 9, 137, 220, 146–151, 156, 159, 160, 171, 177, 180, 181

naturalism *(continued)*
  biohistorical, 6 (*see also* biohistory)
  combinatorial, 203
  emergentist, 29, 41, 214
  expansive, 172–179, 220
  global, 163
  hard, 2, 227n1
  liberal, 9, 137–146, 153, 156, 159, 160, 171, 177, 180, 181, 205, 220
  local, 163, 165, 168, 170, 171
  main features of, 41–46
  materialistic, 221, 41
  mechanistic, 203
  metaphysical, 137, 161
  methodological, 162
  monistic, 8, 9, 41, 43, 46–49, 56–58, 59, 60–66, 70, 80, 114, 120, 137, 140, 153, 156, 159, 171, 182, 215, 219, 224, 233n45
  nonreductive, 19, 29, 41, 49, 215
  object, 151–152, 155
  ontological, 62, 116, 162 (*see also* ontology, naturalistic)
  and panpsychism, 203–205, 214, 217
  philosophical, 71, 89
  physicalist, 203
  pluralistic, 8, 9, 41, 43, 49–58, 59–60, 66, 67–80, 125, 137, 140, 146, 148, 149, 153, 156, 159, 171, 182, 184, 196, 215, 219, 224, 231n14
  pragmatic, 45–46, 137, 151–157, 159, 171, 181, 220, 234n64
  reductive, 29, 99, 146, 163, 172, 215
  religious (*see* religious naturalism)
  religiously informed, 28, 29–30, 39, 102–103, 141, 153
  as research program, 20
  scientific, 137, 138, 139, 141, 161, 171, 172, 176, 181

  soft, 2–3, 8, 227n1
  subject, 151–152, 154, 155–156, 181
  theistic, 15, 16, 17, 160, 174–175
nature, 2–3, 4, 22–23
  as enchanted, 173–174, 175, 216
  as holistic, 216
  laws of, 7, 16, 17, 18, 25, 76, 138, 140, 145, 160, 162, 163, 164, 167, 168, 169, 170–171, 177, 182, 187, 191, 213
  mindfulness of, 105
  as morally ambiguous, 23, 87–88, 125, 126–130, 131, 133, 135
  as Nature, 3, 12–13, 31, 37, 83, 193
  as providing assurance, 126, 132–133
  religion of, 6, 15, 31, 109, 125, 127, 130, 131, 132, 133–134, 135
  as religious object, 2, 3, 13, 22–24, 31, 33, 87, 106, 125, 133–134
  sacredness of, 6, 29, 31, 34, 37, 121, 125, 216
  as Thou, 88, 130, 195–196
neurobiology, 62, 186
neuroscience, 19, 20, 26, 47, 49, 53, 147, 184
New Age spirituality, 15
nihilism, 90, 104–105, 122–124, 161
nociception, 194–195
nonrealism, 37, 124, 153. *See also* realism
  religious nonrealism, 84, 85–86, 92, 94, 134
nonreductionism, 67, 70, 75, 78, 79, 81, 172, 184, 187. *See also* antireductionism; naturalism, nonreductive; reductionism
novelty. *See* ontology, novelty

objectivism, 91, 92, 95, 125–134

Ockham's razor, 44, 160, 168–170, 171
Omega Point, 211, 222, 247n144
ontology
   novelty, 52–53, 67, 68, 69–70, 78, 79, 99, 100, 189
   pluralism, 53, 57, 102, 144
   res extensa, 193
   res potentia, 193
   theistic, 135, 159–180, 207, 214 (*see also* theism, ontological)

pain, 125, 128–129, 147, 150. *See also* evil
panentheism, 9, 17–18, 142, 159–171, 176, 178, 179, 181–182, 220
   and dualism, 163–170
   emergent, 9, 17, 18, 160, 163–167, 170, 176, 179, 220
   panentheistic analogy, 162
   process, 9, 18, 160, 162, 167–168, 176, 179–180
   and religious naturalism, 170–171
panexperientialism, 130, 183, 188–189, 192, 204, 206
panpsychism, 9, 62, 130, 135, 141, 142, 150, 180, 181–217, 245n82, 248n4, 248n5, 248n14
   analogical argument, 186–187
   atheistic, 206–207
   combination problem, 200–203, 208–209, 210, 214–215, 217
   constitutive, 201
   continuity, argument from, 185–186
   and eco-ethics, 216, 217, 222–224
   and emergence theory, 9, 184–185, 189–199
   and naturalism, 203–205, 214, 217
   nontheistic religious, 206–210, 217
   origination argument/ generation problem, 184, 185, 200, 244n12
   participatory, 193
   and religious naturalism, 200, 213–216
   religious relevance of, 206–216
   strong, 187–189, 203, 205
   and subjectivity, 217, 224–225
   and teleology, 212–214, 215, 216, 217, 220–222, 245n82
   and theism, 213–215
   theistic, 210–213, 217
   weak, 187–189, 197, 199, 200, 203, 205–206, 217, 224
pansemiosis, 195, 196, 197–199, 216. *See also* biosemiotics; semiotics
pantheism, 12–13, 29, 34, 71, 167–168, 208, 212, 247n145
perspectivism, 126–127, 128–129, 133
persuasion, 168. *See also* divine action, influence
physicalism, 2, 5, 6, 20–21, 32, 47, 63–64, 90, 109–113, 142, 163, 184–185, 200, 202, 219. *See also* materialism; matter; reality, physical
   epistemological, 48
   nonreductive, 63, 144, 146, 166
   realistic, 36, 183
   reductive, 144
physics, 7, 20, 21, 43, 47, 48, 50–51, 54–56, 57, 59, 62, 79, 102, 114, 115, 163
   measurement event, 192–193, 209
   quantum physics, 25, 192–194, 209–210
pluralism, 30, 57, 152. *See also* naturalism, pluralistic
   emergent, 53, 57, 148
   functional, 154
   semantic, 54–56, 57, 59, 64–66, 80
pragmatic religious realism, 14, 36, 83, 87, 91, 92, 94–106, 107, 109, 116–120, 134–135, 159, 219

pragmatism, 14, 16, 25, 27–28, 32, 34–35, 37, 39, 45–46, 75–76, 94–96, 101, 119–120. *See also* naturalism, pragmatic
problem of competing ontologies, 141–143, 146, 149–150, 160, 177, 180, 181, 182, 220
process philosophy, 87, 186
process theism, 9, 127, 142, 167, 220, 248n5
  and panentheism, 9, 18
protophenomenal properties, 206, 246n116
providence, 17, 164, 165. *See also* divine action
psychological miracles, 160, 164, 165
psychology, 49, 64, 186
purpose, 4, 12, 24, 88, 90, 102, 114, 124, 190–191, 196, 213, 221. *See also* teleology

qualia, 188, 189, 215
quietism, 75, 76, 139, 153, 155

realism, 14, 26, 79–80, 154. *See also* nonrealism
  mental, 183, 184
  metaphysical, 74, 92, 96, 103, 120, 128
  pragmatic, 37–38, 39 (*see also* pragmatic religious realism)
  religious, 8–9, 16, 37, 39, 83–89, 92, 94, 106, 107, 118, 119–124, 135, 219, 230n107
  scientific, 35–36
reality. *See* ontology
reductionism, 19, 47, 55–56, 62–64, 76, 102, 112, 161–163, 187. *See also* antireductionism; naturalism, reductive; nonreductionism
  epistemic, 117–118, 135

monism, 184
  ontological, 142, 169, 201
  psychophysical, 183
relationalism, 126–127
relativism, 86, 91, 126, 128, 129, 173
religion, naturalistically informed, 28–29, 39
religion, indigenous, 15
religion, traditional, 4, 5, 8, 11, 14, 32, 37, 39, 85, 100, 101, 102, 106, 116, 159, 160, 206. *See also* Buddhism; Christianity; Islam; Judaism; religious diversity; theism, Christian
  and religious naturalism, 25–31
  and science, 25–27
religiopoiesis, 104, 106
religious diversity, 85, 86, 113, 134
religious extremism, 96
religious fundamentalism, 31, 96
religious naturalism
  and agnostic naturalism, 148–149
  characteristics of, 11–12
  definitions of, 3–7
  distinctiveness of approach, 14–18
  emergentist-based, 104
  global, 16, 28, 30–31, 39
  and liberal naturalism, 140
  and panentheism, 170–171
  and panpsychism, 200, 213–216
  and pragmatic naturalism, 153
  predecessors, 12–14
  and religious agnosticism, 2, 7, 60, 118, 135, 239
  religious dimension of, 83–107
  and traditional religion, 25–31
religious rightness, 126, 130–132, 133
representationalism, 152, 153, 155
reproductive fitness, 51, 55, 85, 92, 121, 123, 124, 135

Sacred, the, 3, 6, 11, 12, 22, 24, 29, 32, 33, 34, 38, 94, 100, 102–103, 209
salvation, 28–29, 132, 211, 248n14
science, 2, 4, 5, 21, 23–24, 44–45, 53–54, 85–86. *See also* naturalism, scientific
  authority of, 8, 9, 20–21, 38, 42–43, 53–54, 118, 181
  cognitive, 164, 184, 198, 224
  philosophy of, 220
  and traditional religion, 25–27
scientism, 42, 54, 60, 114, 116, 144, 171, 174
secular humanism, 12
secularism, 76, 137
semiotics, 190–191, 194–195, 196, 197, 199, 215. *See also* biosemiotics; pansemiosis
  physiosemiosis, 198
  protosemiosis, 198
sentience, 88, 126, 128, 129–130, 131, 134, 194, 195, 196, 204–205, 240n71
simplicity argument, 240n11
social coherence, 34, 51, 85, 92, 106, 121, 122, 123
soul, 188, 193–194, 209, 224–225
spirit, 12, 15, 89, 211, 222
subjectivity, 61, 62, 64, 115, 147, 185, 189, 200, 201, 202–203, 204, 208, 220
  and panpsychism, 217, 224–225
suffering, 87–88, 109, 125, 127, 128, 129, 131, 132, 134, 135, 212. *See also* evil, waste
supernaturalism, 1, 2, 4, 5, 11, 12, 18, 25–26, 44, 45, 115, 160–161, 172, 204, 206, 237n96. *See also* antisupernaturalism
supervenience, 63–64, 65–66, 112, 142, 144, 145, 146, 240–241n13

teleology, 55–56, 57, 67, 79, 88, 190–191, 196, 208, 209, 211, 223. *See also* purpose
  and "doings," 190–191
  and emergence, 51, 245n82
  and panpsychism, 212–214, 215, 216, 217, 220–222, 245n82
teleonomy, 191
theism, 15–18, 24, 26, 46, 209. *See also* ontology, theistic; panpsychism, theistic; process theism
  Christian, 9, 160, 171–179, 182
  classical, 46, 71, 127, 131, 160, 167, 169, 209
  emergent, 209
  evolutionary, 71
  naturalistic, 6, 15, 16, 17, 160, 167 (*see also* naturalism, theistic)
  ontological, 5, 16–17, 175, 207, 215, 219
  and panpsychism, 213–215
theodicy, 127, 131
transcendence, 11, 12, 13, 17, 44, 86, 93, 98, 119

unexplainability, 63, 65, 67–69, 143
unpredictability, 3, 21, 24, 52, 54, 67–69, 76, 78, 103, 140, 148–149, 166

vitalism, 212–213

waste, 87, 212. *See also* suffering
well-being, 5, 33, 38, 101, 106, 107, 130, 131, 132, 216
wholeness
  personal, 34, 51, 85, 92, 106, 121–122, 123
  spiritual, 28
world. *See also* God, as totality of world
  as God's "body," 17, 160

www.ingramcontent.com/pod-product-compliance
Lightning Source LLC
Chambersburg PA
CBHW030530230426
43665CB00010B/831